P9-DFN-047

THE EVOLUTION OF RADIO ASTRONOMY

THE EVOLUTION OF RADIO ASTRONOMY

by

J. S. Hey

Science History Publications

A DIVISION OF

Neale Watson Academic Publications, Inc.

NEW YORK

First published in the United States in 1973 by
Science History Publications
A DIVISION OF
Neale Watson Academic Publications, Inc.
156 Fifth Avenue, New York 10010

ISBN 0 88202 027 7
Library of Congress Catalog Card Number 73-80636

PRINTED IN GREAT BRITAIN

Contents

Preface

In this book I describe the astonishing advances in astronomy achieved by radio. After nearly half a century with scarcely any awareness of the astronomical potentialities of radio, the subject which had languished in the doldrums for so long suddenly moved forward at a tremendous pace under the stimulus of certain striking discoveries and the impetus of improving techniques. It is curious that a hundred years ago radio waves were completely unknown as a physical entity. In an infinitesimal increment of time, reckoned on any astronomical or terrestrial evolutionary time scale, radio has introduced an entirely new and powerful means of communication, and in radio astronomy has extended its scope to provide a vast range of information about the universe.

The search for this knowledge is a fascinating and compelling task. Radio astronomy is no mere intellectual exercise. The pursuit of its aims presents a story of great challenge and endeavour leading to a deeper understanding of astrophysical phenomena and their cosmological significance. In this book I have sought to blend an account of the aspirations, trials, tribulations and successes of the research scientists who have dedicated their efforts to meet the challenge, with an outline of the many impressive achievements in this important new branch of science.

Acknowledgements

I am greatly indebted to all those who have helped and advised me in discussions, through correspondence, and by sending information and photographs.

I particularly wish to express my appreciation to B. J. Bok, J. G. Bolton, E. G. Bowen, B. F. Burke, A. B. Crawford and the Public Relations Division of Bell Laboratories, R. H. Dicke, V. L. Ginzburg, R. M. Goldstein, O. Hachenberg, D. S. Heeschen, B. G. Hooghoudt, W. E. Howard, F. Hoyle, A. Hunter, P. D. Kalachov, K. I. Kellermann, F. J. Kerr, J. D. Kraus, M. S. Longair, A. C. B. Lovell, D. Lynden-Bell, C. H. Mayer, D. H. Menzel, B. Y. Mills, S. Mitton, J. H. Oort, B. Pagel, Y. Parijsky, G. H. Pettengill, J. Raimond, G. Reber, M. Ryle, G. J. Stanley, H. C. van de Hulst, R. Wielebinski, J. P. Wild.

I wish to thank Sir Richard Woolley and the staff of the Royal Observatory, Herstmonceux for their most helpful assistance and for the provision of library facilities.

In addition to the sources of quotations mentioned in the text, I wish to acknowledge sources of quotations taken from scientific papers as follows: on p. 7, K. G. Jansky, *Proc. IRE*, **23**, 1158 (1935); on pp. 9, 11, G. Reber, *Proc. IRE*, **46**, 15 (1958); on p. 17, G. C. Southworth, *Scientific Monthly*, **82**, 55 (1956); on p. 18, D. W. Heightman, *Wireless World*, **42**, 356 (1938); on p. 83, G. Reber, *Proc. IRE*, **46**, 27 (1958); on p. 87, *Sky and Telescope*, **24**, 132 (1962).

Introduction

Astronomy continually challenges and fascinates the scientist and all who wonder about the mysteries of the heavens and the universe. The story of how radio invaded the realms of astronomy is one of the most remarkable in the history of science. It seemed almost inconceivable that the practical applications of radio could enter so extensively into the academic field of astronomical research, which in past ages has depended solely on optical observation. The book is an account of discovery and research in astronomy by radio methods.

The attraction of scientific research lies in the quest to understand natural phenomena. Its successful outcome brings to the scientist a satisfaction similar to the intellectual and emotional reward experienced by anyone solving a difficult problem. In astronomy there is the added significance of unfolding our knowledge of the universe. In one generation we may fit together only a part of what may seem a never ending jigsaw puzzle. Yet the portion of the picture we resolve appears profoundly important to us, and provides the basis for the next step in discovery and knowledge. Only through understanding can man hope to attain harmony with the universe.

In writing the story of radio astronomy I have become aware of the difficulty of unearthing with any precise certainty the historical facts concerning individual scientists, their discoveries and work. The moment one attempts to examine the facts more closely, even those one has previously accepted as true, the more one feels to be steering a rather shifting course among partial and complex truths. I have no doubt this is why biographies of notable people are so frequently rewritten by different authors with varying emphasis and viewpoints. So far as possible, I have ascertained the accuracy of my account, although a modicum of personal opinion must enter into my interpretation. At least I can say that for some of the early discoveries, I was there. In addition, I have known personally and held discussions or been in communication with most of the research scientists who are mentioned in the book. For the description of scientific results I have over the years waded through much of the vast and formidable sea of radio astronomical papers published in scientific journals throughout the world.

In describing a voyage of discovery it is tempting to point out only the prominent landmarks and forget the continuous coastline linking the more outstanding features. It is necessary to remember that many substantial achievements have been realised through the steady acquisition of knowledge from basic research. In fact, the spectacular discoveries have generally arisen from thorough and persistent investigations with modest objectives made by research workers who, at the same time, have shown a responsive awareness to unexpected possibilities. It is the dedicated explorers who are most likely to stumble on hidden treasures.

The first half of the book is the story of the beginning of radio astronomy, the early discoveries and the launching of research programmes. The second half of the book attempts to survey the subsequent achievements of radio astronomy, to explain how research has progressed, and to assess the significance of this new field of astronomy. My account covers the years up to and including 1970, together with a few salient features of research and planning in 1971 to point the way along paths of new progress. The book describes the evolution of radio astronomy from its hesitant beginning to its emergence as an exuberant and versatile science.

CHAPTER 1

The Beginning of Radio Astronomy

In the years following the first experimental demonstration of radio waves by Hertz in 1888, scientists made several attempts to detect radio waves from the Sun, but all their endeavours were unavailing. It seems extraordinary that 50 years had to elapse before I established in 1942 the occurrence of solar radio emission, paradoxically, when no intentional effort was being made to observe the Sun. Ten years previously, Jansky had unexpectedly recorded radio waves from the Galaxy while investigating radio "atmospherics" producing the crackles and noises that interfere with radio communication. Jansky's discovery in 1932 marked the first successful observation in radio astronomy. It is indeed strange that it took so long to recognise that radio waves were reaching us from celestial sources.

The experiments of Hertz prompted the exciting practical exploration of the properties of radio waves. It was natural that the Sun should have been the first astronomical body to be considered as a possible source of radio emission. Radio covers the long wavelength part of the electromagnetic spectrum extending from millimetres to many thousands of metres. The initial venture to detect solar radio waves was made by Edison, famous for his telegraphic inventions. In 1890, Kennelly who then worked in Edison's laboratory, wrote a letter to the director of the Lick Observatory in California which includes the following extracts (quoted by kind permission of Lick Observatory):

"I may mention that Mr Edison, who does not confine himself to any single line of thought or action, has lately decided on turning a mass of iron ore in New Jersey, that is mined commercially, to account in the direction of research in solar physics. Our time is, of course occupied at the Laboratory in practical work, but on this instance the experiment will be a purely scientific one. . . . Along with the electromagnetic disturbances we receive from the Sun which, of course, you know we recognise as light and heat (I must apologise for stating facts you are so conversant with), it is not unreasonable to suppose there will be disturbances of much longer wavelength. If so, we might translate them into sound. Mr. Edison's plan is to erect on poles round the bulk of the ore, a cable of seven carefully insulated wires, whose final terminals will be brought to a telephone or other apparatus. It is then possible that violent disturbances in the Sun's atmosphere might so disturb either the normal electromagnetic flow of energy we receive, or the normal distribution of magnetic force on this planet, as to bring about an appreciably great change in the flow of magnetic induction embraced by the cable loop.

It occurred to me that, supposing any results were attained indicating solar influence, we should not be able to establish the fact unless we have positive evidence of coincident disturbances in the corona. Perhaps, if you would, you could tell us at what moments such disturbances took place. I must confess I do not know whether sunspot changes enable such disturbances to be precisely recorded or whether you keep any apparatus at work that can record changes in the corona independently of the general illumination. I have no doubt however that you could set us on the right track to determine the times of disturbances optically to compare with the indications of Mr Edison's receiver assuming that it does record as we hope."

There appears to be no account of the actual experiment, which could not have ended in success. It is evident that, firstly, their apparatus was too insensitive, and secondly, at the long wavelengths indicated by the scale of their equipment the ionosphere would have prevented radio waves from the Sun reaching the Earth's surface. Kennelly would realise this twelve years later when, along with Heaviside, he predicted the existence of the reflecting layer in the upper atmosphere.

In 1894 the quest for solar radio emission was continued by Sir Oliver Lodge, who was Professor of Physics at Liverpool University. In a lecture at the Royal Institution in London he said, "I hope to try for long wave radiation from the Sun, filtering out the ordinary well known waves by a blackboard, or other sufficiently opaque substance." Later he wrote, "I did not succeed in this, for a sensitive coherer in an outside shed unprotected by the thick walls of a substantial building cannot be quiet for long. I found its spot of light liable to frequent weak and occasionally violent excursions, and I could not trace any of these to the influence of the Sun. There were evidently too many terrestrial sources of disturbance in a city like Liverpool to make the experiment feasible. I don't know that it might not possibly be successful in some isolated country place, but clearly the arrangement must be highly sensitive in order to succeed."

The "spot of light" refers to his reflecting galvanometer, and from associated descriptions it seems likely that his apparatus was designed to receive at centimetric wavelengths. Although such waves would be able to penetrate the ionosphere, the detection sensitivity of his apparatus was totally inadequate, and in any case Lodge's experiment was ruined by electrical interference generated in Liverpool.

The uncertain knowledge at that time of the influence of the upper atmosphere is illustrated by observations attempted in 1900 by Nordman, a French research student. It is of interest to note that in his doctoral dissertation Nordman refers to previous unsuccessful experiments by Scheiner and Wilsing in Potsdam. Nordman used an aerial 175 m long and set his apparatus up on a glacier at an altitude of 3100 m "to eliminate as much as possible the absorbing action of the atmosphere." With excellent foresight, Nordman predicted that strong outbursts of radio emission would be associated with sunspot activity. He felt so certain of his ground that he blamed the negative results of his experiment entirely on atmospheric absorption. We now know of course that the high altitude was not essential, but that at very long wavelengths he would be defeated by the ionosphere.

A lack of interest followed these early failures to detect radio waves from the Sun. No doubt, it was partly attributable to growing appreciation after 1902 that the ionised reflecting layer in the upper atmosphere was cutting off extra-terrestrial radiation at wavelengths longer than about 20 m. In addition, some workers may have been disheartened by the difficulties that had been experienced. Certainly a lull occurred in the search for radio waves from space. Nevertheless, radio technology was advancing rapidly. In 1921, J. A. Fleming[1] gave a lecture in London entitled "The coming of age of long distance radio telegraphy", a title which aptly denotes the progress of radio in the early 20th century, starting from Marconi's transmissions across the Atlantic in 1901, and culminating with the opportunity in the 1920s for all to "listen in" to "wireless broadcasts". Radio research was by then flourishing in several universities, and in commercial firms like the Bell Telephone Laboratories, USA, and the Marconi Telegraph Company, England.

Radio performance depends not only on the sensitivity of the equipment, but also on the conditions governing the propagation of radio waves through the atmosphere, and on the level of background noise which can be heard in headphones at the receiver. In the 1920s, the limitations imposed by propagational effects and received noise were only partially understood. It was realised that external radio noise could originate from lightning flashes in thunderstorms and the numerous minor discharges of electricity generated in storm clouds, hence the terms "static" and "atmospherics". In England, R. A. Watson-Watt[2] made a comprehensive study of atmospheric radio disturbances at long wavelengths. But it was an investigation in USA of atmospheric radio noise at shorter wavelengths, around 15 m, that led to a discovery which laid the foundation of radio astronomy, as I shall now recount.

KARL JANSKY'S DISCOVERY

In 1930 Karl Guthe Jansky (Plate 1.1), a young physicist on the technical research staff of Bell Telephone Laboratories at Holmdel, New Jersey, USA, was assigned the task of studying the directions of arrival of atmospheric static at wavelengths of about 15 m which were then being deployed for ship to shore and transatlantic communication. For this purpose Jansky planned the construction of a rotatable aerial array 30 m long and 4 m high, providing directional reception of about 30° width in azimuth. The frame was mounted on four wheels taken from

[1] Professor Sir Ambrose Fleming invented the thermionic valve.

[2] Sir Robert Watson-Watt subsequently became famous as the British inventor of radar. He also coined the name 'ionosphere' for the ionised reflecting layers of the upper atmosphere.

Plate 1.1 Karl Guthe Jansky (*by courtesy of the National Radio Astronomy Observatory, USA*)

a model T Ford to allow rotation about a central pivot. With the aid of a motor and chain drive a revolution was completed

Plate 1.2 K. G. Jansky with his rotating array (*by courtesy of Bell Laboratories, USA*)

every twenty minutes, and the contraption, shown in Plate 1.2, was nicknamed the "merry-go-round".

In his 1932 paper describing the investigation Janksy distinguished three distinct types of static: first, the intermittent crashes from local thunderstorms; second, a steadier, weaker static due to the combined effect of many distant storms; third, a steady weak hiss of unknown origin producing a sound in headphones similar to the noise generated within the radio receiver by random thermal agitation of electrons in the components. Initially he attributed the hiss picked up by the aerial to weak manmade interference. But his curiosity was aroused when he realised that the direction of arrival moved round the sky each day. During daylight hours the direction at first seemed to correspond with the Sun, but later he noted a drift in the position of maximum noise which proved that the Sun could not be the source.[1]

Although Jansky's results were expressed simply, they reveal the clear insight of a penetrating mind. His recognition of the third type of static marked the beginning of observational radio astronomy. By following up his initial investigation with a year's recording of data, Jansky clinched the astronomical significance of his discovery. In his second paper, published in 1933, entitled "Electrical disturbances apparently of extraterrestrial origin" he had reached the conclusion that the principal source was maintaining a constant celestial position, the direction traversing the sky with a periodicity of 23 hours 56 minutes characteristic of sidereal objects like the stars. In his paper he explained for the benefit of radio readers the celestial system of coordinates and apparent directions at the Earth. He determined the direction of the main source at approximately right ascension 18 hours, declination 10°S. Speculating on the origin of the radiation, he noticed the near coincidence of this direction with the centre of the Galaxy in the constellation of Sagittarius. It is of interest to mention that Jansky acknowledged the assistance of A.M. Skellett in the astronomical interpretation of the data; Skellett was engaged in ionospheric reflection experiments which subsequently proved to

[1] Jansky was observing during a period of sunspot minimum. We now know that had he been recording during strong solar activity he would almost certainly have also detected radiation from the Sun.

have a special bearing on another aspect of radio applied to astronomy, namely the radar study of meteors.

In a further analysis of the data Jansky in 1935 crystallised his conclusion of the galactic origin of the radio emission. He compared more closely the shape of the graph of intensity on his recording charts with that to be expected, assuming the aerial beam intercepted a source distributed over the Milky Way, which comprises the conglomerate multitude of stars near the galactic plane. His analysis revealed an increase of noise whenever the aerial beam pointed at some part of the Milky Way, the intensity being greatest toward the galactic centre. The most obvious explanation at first appeared to be that the stars themselves were emitting the radio waves, but if so, why was he unable to detect the Sun? To resolve the problem, Jansky sought alternative explanations, and he wrote, ". . . it leads one to speculate as to whether or not the radiations might be caused by some sort of thermal agitation of charged particles. Such particles are found not only in the stars but also in the very considerable amount of interstellar matter that is distributed throughout the Milky Way, which matter, according to Eddington, has an effective temperature of 15 000°C. If the radiations come from such particles one would expect the response obtained to depend upon the directional characteristic and gain of the antenna and the way it is pointed relative to the Milky Way, an expectation which agrees with the observed facts."

Jansky's discovery attracted some publicity, and on 5 May 1933 the *New York Times* carried a front page full column report headed "New Radio Waves traced to Centre of Milky Way". The discovery was also featured in an American radio programme, and the galactic noise received on Jansky's aerial array was broadcast, the commentator announcing "I want you to hear for yourself this radio hiss from the depths of the universe". The listeners' reaction to the ten seconds of broadcast hiss is uncertain; one reporter said it sounded "like steam escaping from a radiator".

It is extraordinary that despite Jansky's papers in scientific journals together with the added publicity, no university or astronomical organisation felt prompted to pursue his surprising discovery. Unfortunately Jansky was not allowed to continue the research further because, from the point of

Plate 1.3 K. G. Jansky indicating the origin of radio noise (*by courtesy of Bell Laboratories, USA*)

view of the Bell Telephone Laboratories, he had fulfilled the practical objective, namely to find the intensity and directional properties of the received noise in order to assess how much radio telegraphy would be affected. A suggestion of Jansky's to build a 100 ft dish-shaped aerial for more detailed investigations was rejected as an unjustifiable expense. The Company considered that any continuation of intensive research on the subject should properly be left to academic research centres. So Jansky, reluctantly, never again worked on radio astronomy. He remained on the staff of BTL, a thorough, methodical and persistent research worker, recognised as an expert on radio noise and interference, and he received an Army–Navy citation for his work on direction finding of enemy transmitters during the second world war.

It is noteworthy to record Jansky's scholastic background. His father was himself a scientist, obtaining graduate and post graduate degrees in Physics and Electrical Engineering at the University of Michigan. There he became a great admirer of Dr. Karl Guthe, and father Jansky decided to name his third son after this physicist and teacher whose guidance had meant so much to him. Karl Guthe Jansky graduated in Physics at the University of Wisconsin where his father became Professor of Electrical engineering. Unfortunately, Karl Jansky suffered from ill health, and for this reason Bell Telephone Laboratories were at first reluctant to offer him the appointment to the post which subsequently enabled him to carry out the brilliant research which laid the foundation of radio astronomy. Although he did not behave like a sick man, his declining health led to his death in 1950 at the early age of 44.

GROTE REBER, THE LONE PIONEER

Radio astronomy would have lapsed into oblivion for a decade but for the inspired initiative of one man, Grote Reber, a young graduate radio engineer of Wheaton, Illinois, USA, who decided to pursue the research as a hobby at his own expense in his spare time. Reber was a great radio enthusiast, who at the age of 15 had built a transmitter receiver and was communicating with radio amateurs throughout the world. After graduation as a radio engineer he worked with a Chicago radio firm. Radio was also his hobby and as he had already achieved his early ambition in radio telegraphy of "working all continents" he now looked for new worlds to conquer. Inspired by reading Jansky's papers he foresaw the fascinating prospect of further exploration if he could meet the challenging problems of devising new techniques of observation. To quote his own words "In my estimation it was obvious that Jansky had made a fundamental and very important discovery. Furthermore, he had exploited it to the limit of his equipment facilities. If greater progress were to be made it would be necessary to construct new and different equipment especially designed to measure the cosmic static". The most crucial questions were, he felt, how to determine the detailed distribution in the sky, and how the radiation intensity depended on wavelength. In the face of prevailing ignorance, he decided to construct a large parabolic reflector with the intention of observing initially at a very short wavelength, about 10 cm. He realised that a parabolic reflector would have the advantage of providing a narrow, symmetrical beam and would also enable the wavelength to be altered simply by changing the receptor at the focus. In the choice of operating wavelength, Reber was guided by two considerations: he could achieve better angular resolution, and also, he argued, if Planck's thermal emission law applied, the radiation should be stronger at shorter wavelengths.

With this well conceived plan, Reber began to build the first radio telescope specifically designed for radio astronomical observations, a mammoth practical task for one man to undertake. He would have preferred a full steerable mounting but this was far too expensive, so he decided on a meridian transit instrument steerable in elevation only, and relying on the Earth's rotation to scan the heavens. The metal parabolic mirror was to be made as large as possible consistent with available funds. Even the lowest estimates from outside contractors were prohibitive. Only one course seemed open to Reber; he must design and build the telescope himself in his own back garden. The prime consideration in design was to maintain a stable configuration of the reflector and focal mount at all elevations. The reflector had to be as large as possible, and yet deviations in shape must not exceed a fraction of a wavelength. Balancing the cost of materials against the structural demands, Reber finalised the parameters of his design and decided on a sheet metal surface of 31 ft diameter, to be mounted on a wooden supporting structure for the sake of cheapness and ease of construction. The reflector surface consisted of 45 pieces of 26 gauge galvanised iron sheet screwed on 72 radial wooden rafters cut to parabolic shape. Reber cut, drilled and painted all the parts; and except for part-time assistance of two men on foundations and erections, Reber personally put together the radio telescope piece by piece, and completed the entire job in four months from June to September 1937. The building of the telescope cost Reber 1300 dollars, a princely sum in 1937 dollar values, but neither cost nor effort deterred his faith in this individual enterprise. The neighbours viewed the rising structure with amazement. When it was completed, Reber notes "The mirror usually emitted snapping, popping and banging sounds every morning and evening . . . due to unequal expansion in the reflector skin. When parked in the vertical position great volumes of water poured through the centre hole during a rain storm. This caused rumours among the local inhabitants that the machine was for collecting water and for controlling the weather. . . "

After an experimental study of the sensitivity of radio receivers at centimetric wavelengths he chose a crystal detector followed by an audio triode amplifier. His initial attempts, at $\lambda = 9$ cm, to observe the Sun, Moon, planets, and brightest stars produced no response. Undeterred, he changed to $\lambda =$ 33 cm, a wavelength for which an acorn triode proved the most suitable detector. Reber writes "During the autumn of 1938 and during the winter a variety of observations, both by day and night, were made with both polarisations. All the same objects were examined again without any positive results". So he concluded that Planck's blackbody law was not obeyed by the extraterrestrial noise, and in 1939 we find him starting again with a new receiver at $\lambda = 1 \cdot 87$ m and successful at last in detecting radio emission from the Milky Way despite increasing trouble from ignition interference at this longer wavelength. At this time Reber was recording measurements from midnight to 6 AM while the manmade interference level was at a minimum. Then after breakfast he drove 30 miles to Chicago where he earned his living by designing receivers for a radio company, and back again in the evening, sleeping after supper until midnight when he resumed observations.

Reber's radio telescope, shown in Plate 1.4, and his persistence in overcoming technical difficulties, rank him as a great pioneer of radio astronomy. His research culminated in important contributions which revived interest in this neglected subject.

Let us now look at some of the results of his work. Reber, in his first paper in 1940 entitled "Cosmic static", determined the radiation intensity at $\lambda = 1 \cdot 87$ m and confirmed the source distribution as lying predominantly along the Milky Way. He then made an important step in theoretical interpretation by evaluating the intensity of radio emission from free electrons

Plate 1.4 G. Reber's radio telescope at Wheaton, Illinois (*by courtesy of G. Reber*)

during encounters with positive ions of ionised hydrogen in interstellar space. The formula derived in a different context by Kramers in 1923 to explain the continuous X-ray spectrum from metals bombarded by fast electrons had previously been applied to optical radiation from stars and interstellar gas by Eddington. Reber showed that the radio emission agreed reasonably well with the expected radiation from interstellar electrons as indicated by Kramers' formula.

The problem now at last attracted the attention of astronomers, and in the *Astrophysical Journal*, 1940, Henyey and Keenan developed the theoretical treatment more fully. They showed that the ionised gas should be partially transparent at shorter wavelengths, giving an almost constant power flux, but at wavelengths greater than about 5 m it should become opaque and emit with the intensity according to Planck's blackbody law. The agreement with Reber's measurements at about 2 m wavelength was satisfactory. But the intensity recorded by Jansky at about 15 m represented a radiation temperature of about 150 000 K, and yet it was known that the electron temperature of ionised hydrogen could not greatly exceed 10 000 K A major discrepancy was now exposed; could there be a different radiation mechanism at long wavelengths? As we shall see later,

Plate 1.5 Grote Reber at NRAO in 1960 (*by courtesy of the National Radio Astronomy Observatory, USA*)

a decade elapsed before an answer could be found, after much new light had been thrown on the problem by fresh discoveries.

A further achievement by Reber was his production of the first radio maps of the Milky Way. The beamwidth of his radio telescope, about 12° at $\lambda \sim 2$ m, enabled him to draw a contour map of the distribution of radio noise which showed its relation to the Galaxy, the structure of the main peak at the galactic centre in Sagittarius, and the subsidiary peaks in Cygnus and Cassiopeia. (Reber's map is shown in Figure 6.3(a) on p. 132). Attempts to detect individual objects, like planets, stars, and nebulae, were unavailing, but his 1944 paper reported detection of the Sun. In the meantime, however, observations of solar radio waves had already been made elsewhere, and I will now return to the question of the Sun, which for 50 years had eluded all attempts to detect it by radio.

RADIO DETECTION OF THE SUN

I must now digress to write about myself and how I became directly involved during February 1942 in the discovery of a new and exciting solar radio phenomenon. One should remember that 1942 was wartime; my experience of intense radio emission from the Sun was the consequence of a chain of events connected with my investigations of the performance of British Army radar. I was a physicist, but my knowledge of radio was negligible until I received six weeks intensive training from J. A. Ratcliffe[1] at the Army Radio School at Petersham, Surrey, after which I was drafted to the Army Operational Research Group under B. F. J. Schonland[2]. I was fortunate there in being associated with a group of distinguished research scientists who were tackling a variety of Army problems in the aura of objective scientific research. At the onset of the war the newly invented radar systems were subject to seemingly temperamental errors. I was allocated the task of advising on Army radar performance, and although I felt it could not be true, I was generally regarded as a radio expert. Curiously enough,

[1] J. A. Ratcliffe, FRS, a Cambridge radio physicist, later became Director of the Radio and Space Research Station, Slough.

[2] Sir Basil Schonland, FRS, Professor of Geophysics, Witwatersrand, S. Africa, later became Director of the Atomic Energy Research Establishment, Harwell.

buoyed by the esteem of eminent colleagues, combined with the successful solution of a series of problems, I appeared to fulfil the reputation bestowed upon me. The appeal of the new subject, the sound basic training, the urgent demands of defence, and the invigorating research environment, all fed my enthusiasm.

During 1941, the enemy made increasing endeavours to jam radar operation. The War Office became anxious lest their radar devices, particularly vulnerable to airborne jamming, might be rendered useless. On 12 February, 1942, the passage of the German warships Scharnhorst and Gneisenau through the English Channel, slipping by almost unnoticed until it was too late to muster any effective attack on them, to the accompaniment of radar jamming from the French coast, resulted in a drastic reappraisal of the jamming menace. The War Office decided to augment its efforts to contend with radar jamming and sought the assistance of the Army Operational Research Group in dealing with this awkward problem. The investigation of jamming is hardly an appealing subject for a scientist, on the face of it a negative, irksome task. Nevertheless, the challenge had to be met and I readily accepted the invitation to be responsible for analysing Army radar jamming, and advising on anti-jamming measures. I cooperated with Army staff officers in devising instructions for radar operators and organising an immediate reporting system. A mobile J-watch laboratory was strategically sited on the cliffs at Dover and manned by a member of my team. I had a peculiar role as a civilian scientist holding a key position in an Army system and the work proved not dull but exciting with my advice often being sought urgently by Anti-Aircraft Command and by the War Office.

On 27 and 28 February 1942, a remarkable series of reports from sites in many parts of the country described the daytime occurrence of severe noise jamming experienced by anti-aircraft radar working at wavelengths between 4 and 8 m, and of such intensity as to render radar operation impossible. Fortunately no air raids were in progress but alarm was widespread at the incidence of this new form of jamming, all wondering what it might portend. Recognising that the directions of maximum interference recorded by the operators appeared to follow the Sun, I immediately telephoned the Royal Observatory at Greenwich to inquire whether there was any unusual solar activity, and was informed that although we were within two years of the minimum of the sunspot cycle, an exceptionally active sunspot was in transit across the solar disk and situated on the central meridian on 28 February. It was clear to me that the Sun must be radiating electromagnetic waves directly—for how else could the coincidence in direction be explained—and that the active sunspot region was the likely source. I knew that magnetron valves generated centimetric radio waves from the motion of electrons in kilogauss magnetic fields, and I thought why should it not be possible for a sunspot region, with its vast reservoir of energy and known emission of corpuscular streams of ions and electrons in magnetic fields of the order of 100 gauss, to generate metre wave radiation.

When I wrote a paper giving details of the event, my director B. F. J. Schonland recalled Jansky's discovery of galactic radio noise with which I had been hitherto unacquainted. What was surprising, however, was that several radio scientists, experienced in ionospheric and communications research, were sceptical of my conclusions. They found it hard to believe that such powerful radio outbursts had escaped notice in previous decades of radio research. It seemed almost an effrontery for a comparative novice in the field to be presenting a paper on an energetic solar radio phenomenon.

The discovery of the intense radio emission from the Sun had some features in common with Jansky's discovery of cosmic radio noise. Both were examples of observations for one purpose leading to hitherto unknown phenomena. In both instances, the aim had been to study types of interference limiting the effectiveness of practical systems.

Later in 1942, a different kind of solar radio measurement was made independently at the Bell Telephone Laboratories, USA, by G. C. Southworth, who had been designing sensitive centimetric receivers of low inherent noise level. He realised that the optical and infrared radiation from the Sun followed Planck's Law for a full radiator at 6000 K, the photospheric temperature of the Sun. If the Sun normally radiated according to the same law in the radio tail of the spectrum, then Southworth thought he might detect the radiation at centimetric wavelengths. On completing the low noise microwave receiver at $\lambda = 3{\cdot}2$ cm, Southworth states ". . . it was natural that we should first point the antenna at the Sun. This was first done at my request by one of my associates, A. P. King, on 29 June 1941. We found as expected, that the solar noise represented a small increase in the total noise output".

Southworth, in his initial report, concluded that the results confirmed that the quiet Sun behaved as a blackbody radiator at 6000 K. Like others, he had not appreciated that the bulk of the microwave radiation arises from the much hotter highly-ionised chromosphere which lies above the photosphere, or he would not have so readily accepted his first estimate. Subsequently, he re-examined his data, and on recalculation the value of the microwave brightness temperature was revised to 20 000 K.

The obstacles which had so long barred radio observation of the Sun had at last been removed by my paper and that by Southworth. For both of us, security regulations prohibited general circulation of our 1942 papers and republication of the results in scientific journals had to await the end of hostilities.

After the war, in February 1946, a giant sunspot attaining an area of 1/200 of the solar hemisphere was found to be radiating strongly, and provided the opportunity for a detailed investigation by my team at AORG. We discerned two forms of the abnormal solar radio emission, the long lasting noise storm associated with the sunspot, and the intense bursts accompanying the solar flares which occur at intervals in the vicinity of active sunspots.

The previous failures of other workers to recognise abnormal solar emission illustrates, I think, the stultifying effect of clinging to established viewpoints, in this case biased by the early negative attempts. The intense noise is so strong that it had almost clamoured to be observed in the past, but the recognition of the spasmodic outbursts had been missed. Looking back, there had been many reports of high-level noise on short-wave receivers during solar activity, but their true significance had not been appreciated. Even in astronomical papers, for instance, H. W. Newton in 1936 referred to the "radio fizzlies" reported on short-wave communication links, preceding the fade-outs known to accompany strong solar flares. In 1938, an amateur, D. W. Heightman, came near to the correct explanation when he wrote "At such times (when fade-outs occur) the writer has often observed the reception of a peculiar radiation, mostly on frequencies over 20 Mc/s which on the receiver takes the form of a smooth though loud hissing sound. This is presumably caused by the arrival of charged particles from the Sun on the aerial". Two Japanese research workers, Nakagami and Miya, were within an ace of finding the true nature of the radiation in 1939 when they measured the direction of arrival, including elevation, of the noise at wavelengths of 23 m and 17 m, and although the direction corresponded with that of the Sun, they concluded that the noise probably originated in or near the E layer of the ionosphere.

The missed discovery undoubtedly astounded a number of experienced radio scientists. I well remember Sir Edward Appleton's astonishment at a meeting in 1945 when I remarked that I was contemplating publishing in a scientific journal my 1942 paper on solar radio emission for, by some mischance, no one had informed him of the 1942 episode. Appleton then recalled the "hiss" mentioned in the past by radio amateurs during solar activity. Appleton[1] was at that time essentially an administrator at the head of the Department of Scientific and Industrial Research, but after his brilliant career in radio science he naturally retained a passionate concern for radio. He found it hard to limit himself to a vicarious interest in radio research, and he now saw an opportunity for collaboration and sharing in an exciting new field. Such a liaison brought the advantage of many friendly and informative discussions, for he frequently telephoned me at home in the evenings to enquire about the AORG observations, to discuss their implications and to propose joint papers in scientific journals. But at no time had he any part in the direction of the team or the experimental programme and only once visited the site and equipment, on the occasion of a press visit. With reluctance I became aware of the pitfalls of an undefined association. Appleton's incursion into our research findings began to arouse discontent within my own team and the liaison could not survive. In retrospect, one can understand the feelings of scientists promoted to senior administrative posts and their unwillingness to

[1] Sir Edward Appleton, FRS, Nobel Laureate, previously Professor of Physics at King's College, London, and at Cambridge.

surrender to newcomers their supremacy in scientific domains. One must also admit that rivalry is no more an uncommon failing among scientists than in other walks of life and in recompense frequently acts as a stimulant to endeavour.

LONG RANGE ROCKETS AND METEORS

A surprising sequence of events guided me into other aspects of radio astronomy and culminated in two further important discoveries. Operational radar research in wartime brought the most stimulating phase of my scientific career. The exigency of the military situation tremendously quickened the pace of research. It was essential to get to the roots of a problem. If time pressed and an ad hoc solution had to be proposed, then supplementary theory and subsidiary experiments were rapidly brought to bear to test its validity. The mutual helpfulness and ready sharing of knowledge by fellow workers were conducive to rapid progress. The work was hard, but rewarding in making vital and worthwhile contributions to the national war effort, and scientifically satisfying especially when it led to a fresh understanding of radio physics and to new discoveries.

There was an unending succession of problems. At the risk of appearing egotistical let me continue to recall certain events in which I was involved because they have relevance to radio astronomy. I am cognisant that my contribution constituted nothing more than a small part of the operational situations here mentioned. In 1943, no sooner had I overcome the difficulty due to ground reflections in the radar tracking of V1 flying bombs at very low heights by placing diffracting screens[1] in front of the anti-aircraft radars, than the imminent threat of bombardment by V2 long range rockets demanded a solution to the problem of radar observation at great heights and distances. A high elevation RAF radar system for watching approaching rockets was too far from completion and an alternative had to be sought. Sir Robert Watson-Watt was chairman of the interservice Crossbow Radar Committee which had been allotted the task of deciding radar policy. At a meeting

[1] The method was published in 1956 (J. S. Hey and S. J. Parsons, *Proc. Phys. Soc.*, **B 69**, 321) in a paper entitled 'The radar measurement of low angles of elevation'.

of the Committee I suggested that if Army 5 m wavelength radar could be fitted with large aerials directed at high elevation they would have adequate performance to detect V2s over part of their trajectory as they descended towards Britain. The scheme was readily adopted and in six weeks a chain of modified AA radar was installed around the coast just before the V2 attack was unleashed. Our record during the first week was deplorable; the only V2 detected was a wild shot that nearly hit the watching radar. Installation faults were hastily diagnosed and remedied, and afterwards every V2 was successfully detected. The system was subsequently developed to predict the point of fall, and when the V2 bombardment ceased, AA Command were preparing to shoot a barrage at the rockets in an attempt to explode them in mid air.

The capability of the radars, looking at elevations of about 45° to detect rockets at ranges of 100 miles and more, was strained to the limit, and attempts were made to improve sensitivity by introducing specially designed receivers. Curiously, no improvement in detection range ensued, and it was only after J. M. C. Scott[1] reminded me of the cosmic noise discovered by Jansky did I realise and verify that the limiting factor was indeed the external noise from the Galaxy.

Another problem arose from spasmodic transient echoes at heights around 60 miles. These echoes, appearing at an average rate of 5 to 10 per hour, were a maddening nuisance and caused frequent false alarms. The existence of two forms of irregular ionisation in the ionospheric E-layer was already known, very brief "short scatter" and the longer lasting "sporadic-E". As long ago as 1932 in USA, Skellett, and Schafer and Goodall, had demonstrated that meteors could produce sudden increases of irregular ionisation as they burned up in the upper atmosphere. The distinction between "sporadic-E" and "short scatter" and their possible variety of causes long remained uncertain and confused, but it gradually became clear that short scatter echoes could be distinguished by their occurrence at comparatively short wavelengths, of only a few metres. It was evident that the V2 radars were observing short scatter echoes. Appleton favoured the meteoric hypothesis of their origin, but Eckersley

[1] J. M. C. Scott is a Lecturer in Physics at Cambridge.

who had studied the transient echoes in great detail was disinclined to accept this opinion.

At the end of the war in 1945 there were no immediate tasks in view for AORG. Some members of the group dispersed at once to former posts at universities and elsewhere, but others, like myself, remained to await posssible developments in government research. I grasped the opportunity in this interim period to delve more deeply into the wealth of phenomena we had uncovered in our operational research. With an excellent well balanced team, S. J. Parsons (electrical and mechanical engineer), J. W. Phillips (mathematician), G. S. Stewart (electrical engineer) and myself (physicist) as leader, we embarked on projects utilising modified radar for radio astronomy research. Initially, we were also helped by AA Command who were magnanimous in their gratitude to AORG whose various sections had advised in many ways such as gunnery, training and so on, as well as radar.[1] The enlightened attitude towards research, both by the Ministry of Supply and by the Services was gratifying. In a close involvement with Army personnel, their open friendly interest was a revelation and I think a characteristic of their profession with its very high standards, its readiness to face problems, and its sense of mission.

Having accepted the offer of AA Command's assistance, I conceived a simple experiment to decipher the character of the short scatter echoes. This we put into effect in Spring 1945, for the V2 attack had then ceased. We organised a combined watch on a single zone of the E-layer from three widely separated sites, shown in Figure 1.1, comprising two V2 radar tracking sites, at Aldeburgh and Walmer, and a similar radar at the AORG experimental site in Richmond Park. If the short scatter echoes were seen simultaneously at all sites, we should have a very precise location of position and movement, from the three measurements of range. Alternatively, I argued, if the echoes were sensitive to aspect, for instance, if they were long, narrow, ionised trails of meteors, this basic characteristic would

[1] In his despatch to the War Office, General Sir Frederick Pile stated: "While full credit must be given to troops of all kinds, and indeed their conduct under very hazardous and trying conditions was beyond all praise, the foundation of success, however, was laid by the scientist, both civilian and in uniform. The Operational Research Group has already been referred to. Its work was brilliant". (Quoted from the Supplement to the London Gazette, 16 December, 1947)

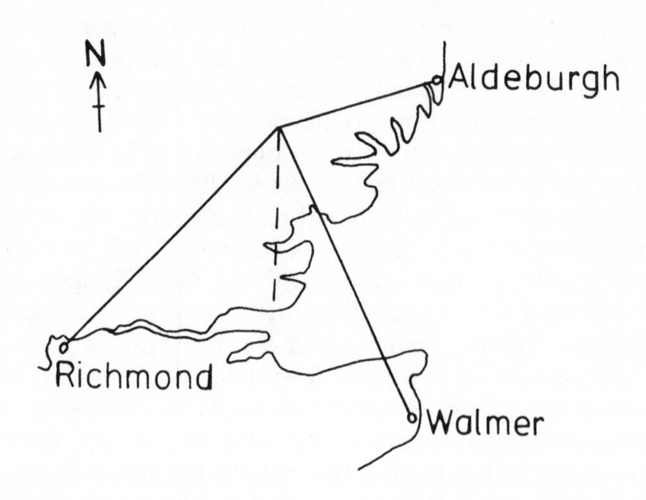

Figure 1.1 The radar meteor watch planned by Hey and Stewart in 1945.

be revealed by the three different views of the same trail. The results were decisive; the echoes never appeared simultaneously on the three radars, so it was obvious that the echoes must be highly sensitive to aspect. It therefore seemed likely that the echoes were produced by meteor trails, long columns of ionisation that would reflect back like mirrors when viewed at right angles. We next found a diurnal variation in the numbers of echoes, a peak rate of occurrence appearing first at one site and a few hours later at another site. We hit at once on the explanation, that in addition to a background of widely distributed meteors, there were predominant streams in particular directions. As the Earth rotates, first one radar beam would intercept a stream at right angles, and a few hours later another radar beam would move into the direction of maximum reflection. By noting the times of the peak rates of occurrence of echoes at the different sites we were able to deduce the celestial directions, known as meteor radiants, of the main meteor streams. With the aid of radar we were now for the first time "watching" daytime as well as night-time meteor streams.

After June 1945, AA Command's help dwindled with the

demobilisation of the forces, so Stewart and I continued the meteor echo research with our Richmond Park radar, carrying out one experiment after another to clinch the meteoric nature of the short-scatter echoes. High rates of occurrence of echoes accompanied the well-known meteor showers, and combined visual watch revealed many exact coincidences between radar echoes and visible meteors. To my mind, two key factors had enabled us to unravel the uncertainty that had for so long clouded the interpretation of the short scatter echoes, namely, the technical advances in precision and sensitivity of radar equipment, and the keen research environment. Curiously enough, once more, a phenomenon impinging itself as a hindrance to the primary function of the radar apparatus had led to important astronomical deductions.

In addition to our research at AORG on solar radio noise, and on meteors, we also embarked on a project to make a detailed map of cosmic radio noise using a modified radar tracking receiver in Richmond Park. This map of the radio distribution, the first at 5 m wavelength had much similarity with that produced by Reber at shorter wavelengths. As I shall describe later our work on cosmic radio noise led us to yet another startling discovery, a discrete source in the sky, which proved to be the first of a vast multitude of sources, the starting point of a fascinating era in astronomical research. This discovery and other developments in radio astronomy at AORG will be discussed later in relation to the rapidly growing research centres elsewhere. Now let us look at a few more basic steps taken by other groups which further contributed to the full appreciation of the potentialities of radio astronomy.

THE PREDICTION OF THE HYDROGEN LINE

Although Holland was under military occupation during the war, copies of the *Astrophysical Journal*, published in USA, reached the Observatory at Leiden. Papers by Reber appeared in the Journal in 1940 and 1944, the latter showing his radio map of the Galaxy. J. H. Oort, the director of the Leiden Observatory was greatly interested in Reber's articles; and at a colloquim in 1944 discussing Reber's observations of continuum radiation, Oort pointed out the advantages that would accrue if monochromatic lines existed in the radio spectrum. A member of the observatory research staff, H. C. van de Hulst, after a detailed consideration, reached the conclusion that a radio line at 21 cm wavelength might be observable from neutral atomic

Plate 1.6 Van de Hulst reading his paper on the 21 cm hydrogen line. (This photograph taken in 1955 is a reconstruction of the 1944 meeting). (*By courtesy of H. C. van de Hulst, Leiden*)

hydrogen in interstellar space. On the face of it, the hyperfine transition scarcely seemed a hopeful proposition, for the temperature of interstellar atomic hydrogen would be very low, and the hyperfine transition probability only about once in millions of years, while the density of hydrogen would not average more than about 1 atom per cubic centimetre. But realising that the vast extent of the Galaxy compensates for these discouraging factors, van de Hulst in 1945 suggested that a search should be made. As had been foreseen, the prediction of the 21 cm line proved to be of great fundamental importance, a basic step in establishing a valuable method of studying the structure of the Galaxy.

RADAR ECHOES FROM THE MOON

Radar detection of the Moon presented an obvious test of radar capability. To obtain echoes from the Moon at this time could be regarded as a "tour de force" of radar power and sensitivity. Although the Moon has a large angular size, it lay beyond the range of current radar equipment. The echo return from a target falls off inversely as the fourth power of the range and so presents a serious limitation to radar's ability to reach out to distant celestial objects. Nevertheless, by the end of the war, with the aid of specially designed systems, the possibility of obtaining lunar radar echoes had at last come within grasp.

Two experiments, poles apart in method of execution, were successful in detecting radio reflections from the Moon early in 1946. One of the experiments, conducted by the United States Army Signal Corps Laboratory, was initiated by Lt. Col. J. H. De Witt as an immediate post war opportunity to achieve a goal he had long cherished. The experiment was named Project Diana, and in August 1945, the modification of Army radio equipment commenced in order to attain the objective. An aerial array rotatable in azimuth, and consisting of 64 dipoles at 2·7 m wavelength utilised ground reflection in order to increase the gain of the aerial beam, which was fixed at low elevation. A CW transmitter was keyed to transmit $\frac{1}{4}$ second pulses, and the receiver bandwidth was narrowed for optimum exclusion of unwanted noise. First echoes were recorded on the cathode ray tube display on 10 January, 1946. The distance could only be estimated to the nearest 1000 miles, but at least a beginning in lunar radar measurements had been made.

The other experiment, a most creditable effort by Z. Bay in Hungary, achieved its success through an ingenious and unconventional approach by his method of integrating the receiver output over long intervals by means of chemical electrolysis. The output was connected in turn to a series of water voltameters, each corresponding to a different time interval after the transmitted pulse. The procedure was continued for periods of half an hour. In this way, he demonstrated that one of the voltameters, the one set at $2\frac{1}{2}$ seconds delay, the time taken for the radio pulse to travel to the Moon and back, liberated the most hydrogen, proving that an echo from the Moon was being recorded.

RADIO EMISSION FROM THE MOON

To detect radio waves emitted from the Moon was obviously a question of designing a sufficiently sensitive receiver. Infrared devices were capable of measuring the thermal radiation from the Moon; in comparison radio suffers from the very small thermal power emitted at radio wavelengths. It is convenient to express power in terms of equivalent temperature; the noise generated internally in centimetric receivers currently available in 1945 would, had it been produced thermally, correspond to about 7500 K. So it was by no means easy to detect the Moon, with its mean temperature of about 230 K, especially as the amplification of a straight receiver is liable to fluctuate. Furthermore, if the aerial beam were larger than the Moon, the effective temperature of the received radiation would be correspondingly diluted. For instance, a beam of solid angle ten times that of the Moon would result in a mean aerial temperature of 23 K. To measure alterations in temperature with the changing phase of the Moon, one would wish to be able to record changes of 1° or less. How was such sensitivity to be achieved?

Provided a receiver is sufficiently stable, sensitivity can be improved by collecting radiation over a wide bandwidth, and then averaging the receiver output for as long as possible, the process known as integration. This method was used by Southworth in his centimetric observations of the Sun; but variations in amplification set a limit to the sensitivity that can be achieved in this way with a conventional "straight" receiver. The difficulty was overcome by R. H. Dicke in a receiving system he designed when working at the Massachusetts

Institute of Technology (MIT) on a wartime project to investigate atmospheric absorption at short wavelengths. In order to determine the radiation coming from the atmosphere he required a very sensitive radio receiver. For this purpose, he devised a rapid switching method so that the incoming signal was repeatedly compared with a thermostatically controlled standard source. At the output a switch-frequency rectifier measured the difference. Since both the celestial signal and the standard noise signal were subjected to the same amplification, the gain fluctuation of the receiver became relatively unimportant. With an input bandwidth of 8 MHz and an output time constant of 2 seconds Dicke demonstrated that he could detect intensity changes corresponding to less than half a degree of temperature.

It was natural that Dicke should look at the Sun and Moon with his sensitive radiometer, and so, observing at 1·25 cm wavelength with a parabolic reflector of 18 inches diameter, Dicke and Beringer measured the temperature of the nearly full Moon as 292 K, and the quiet Sun 11 000 K. A considerable allowance had to be made for atmospheric absorption at this wavelength, and for the beamwidth which covered 2° to half power; taking these and other factors into account they estimated their probable error at ± 10%.

Incidental to their measurements of atmospheric radiation and absorption, Dicke and his colleagues derived an upper limit of 20 K for the residual temperature of the celestial background which they called "the radiation from cosmic matter". Their evaluation of the upper limit was inevitably rather crude because of the difficulty in making accurate allowance for the other extraneous sources of noise both from the atmosphere and equipment. Nevertheless, it was particularly interesting as an endeavour to measure background radiation, a subject which was subsequently to prove of vital consequence in the understanding of the past history of the universe. Some twenty years or more later at Princeton University, Dicke became deeply interested in the background radiation and its cosmic significance. Plate 1.7 shows Dicke and his colleagues in 1945 preparing to make observations with the radiometer connected to a horn aerial. In the picture, the "shaggy dog" which Dicke is holding in front of the horn is a piece of absorbing material.

Plate 1.7 Preparing to make measurements with the Dicke radiometer in 1945. Left to right: R. L. Kyhl, E. R. Beringer, A. B. Vane, R. H. Dicke (*by courtesy of R. H. Dicke, Princeton University*)

The great merit of Dicke's technique was the attainment of such acute radio sensitivity, and the method afterwards came to be almost universally applied in one form or another in making radio astronomical measurements.

All the work so far described was well under way prior to or during 1945, although some of the published papers describing results are dated 1946. By now the foundations had been truly laid both in discoveries and techniques for the realisation of the potentialities of radio methods in astronomy. The next chapter describes how a new wave of research became organised to herald a full tidal flow of progress into the exploration of space by radio.

CHAPTER 2

The Rise of Radio Astronomy

In the autumn of 1945, university professors and lecturers who had been engaged in university research were returning to their peacetime posts. The war years had accelerated progress in scientific and technological fields such as radar and nuclear physics, and the influence of advances in techniques was soon to be felt in almost every branch of research. Electronics became a commonplace adjunct of many kinds of experimental investigation. Research by individual scientists with primitive apparatus gave way to concerted efforts by teams with sophisticated equipment. Scientists resuming their academic careers prospected among the resources of invention and discovery uncovered during the transitional period of hostilities to find what could fruitfully be incorporated in their peacetime research programmes. Two physicists who had been engaged on airborne radar research at the Telecommunications Research Establishment (TRE) visited AORG to learn more about our experimental results and to see our equipment in Richmond Park. One of them, J. A. Ratcliffe, was now returning to Cambridge to resume direction of radio and ionospheric research at the Cavendish Laboratory. The other, A. C. B. Lovell, returning to cosmic ray research at Manchester University, was seeking a method of observing radar echoes from cosmic ray showers. Their own research objectives, together with the ideas stimulated by the AORG observations, subsequently led, through a sequence of events, to the inauguration of the great radio astronomy centres at Cambridge and Manchester. At the same time, another major research institution was initiated in Australia. I will now describe separately the stages in the development of these three groups.

HOW JODRELL BANK BEGAN

At Manchester University, Lovell held the post of lecturer in Professor Blackett's Physics Department where the research was primarily devoted to studies of cosmic rays. Lovell's wartime radar work at TRE spurred his imagination to a consideration of the feasibility of observing radar echoes from cosmic ray showers. He discussed the idea with Blackett in 1941, and although both were preoccupied with wartime problems, they published a short joint paper in the Proceedings of the Royal Society on the likelihood of detecting radio reflections from the columns of ionisation produced in the atmosphere by the passage of cosmic ray showers. On returning to University research in 1945 Lovell immediately planned to put the idea to experimental test. He concluded that the AORG equipment, the type that had been designed for watching V2 rockets and which we had adapted for meteor echo investigations, was ideally suited to his purpose. So Blackett and Lovell negotiated with the War Office and Ministry of Supply for the transference of an identical equipment to the University, and it was agreed that I should go with my colleague S. J. Parsons to the University for two or three weeks to set up the radar in optimum working condition, to fit a brightness-modulated display on the cathode ray tube for cinerecording of transient echoes, and to advise on the operation of the equipment. The radar was duly installed in a quadrangle near the Physics Laboratory. Our visit proved quite hectic, interspersed by numerous discussions with Blackett, Lovell, and other senior members of the Physics staff. Lovell was bubbling over with enthusiasm, and bombarded us with questions and ideas, so we were frequently engaged in an amiable battle of argument, as to whether this, that or the other of Lovell's latest suggestions was or was not feasible. From time to time there emerged ideas we all approved, and so Lovell's provisional plans progressed in excited anticipation. But we foresaw two serious and perhaps overwhelming snags. One was the ever-present interference, for the University is situated near the heart of Manchester, and a conglomeration of electrical radiation emanated from the industrial and domestic surroundings, further heightened by the screaming surge of spikes and impulses from the electric trams running past the

University. To escape from interference a new site in a rural area seemed imperative. Even so, the cosmic ray echo search appeared a doubtful proposition, for the estimated duration of ionisation was far shorter than the length of the radar pulse, a few microseconds transmitted at millisecond intervals. At the outset I felt there was only a remote prospect of catching cosmic ray echoes. When Parsons and I left Manchester we thought it highly probable that Lovell would abandon the elusive cosmic ray echoes, and direct his attention instead to the readily observable meteor echoes.

Our prediction proved correct, for by early 1946 Lovell had moved out to a site in the Cheshire countryside where the botanical department of the University had an experimental horticultural site at Jodrell Bank; here, after unavailing efforts to detect echoes from cosmic ray showers, Lovell began to pursue research on meteor echoes with great vigour. During occasional visits, I have vivid recollections of Lovell's tremendous energy, and of drives in his open sports car as he chased between University and Jodrell Bank while he enthusiastically propounded his research plans. There was clearly no room for stragglers among his team at Jodrell Bank, and soon he was pouring out a succession of research papers, keenly pursuing the lines of work I had initiated at AORG.

METEORS AND THE GIACOBINID SHOWER

For my group and Lovell's, the great Giacobinid shower of 1946 came as a climax, like a huge firework display to acclaim the successes we both had achieved in radar meteor astronomy. Astronomers had predicted that the Giacobinid shower due to appear on the night of 9 October 1946 would be one of the most spectacular meteor displays of the century. Stewart and I had only recently published our proof that the scatter echoes came from meteor trails and shown how we could determine the radiant directions of the showers. We were agog with keen anticipation knowing, with 100% confidence, that if the Giacobinid shower turned up with the predicted intensity, our radar recorders would be crowded with echoes. The Ministry of Supply, who sponsored the research at AORG, decided to invite the Press to view our radar installation in Richmond Park on 9th October. Appleton was somewhat disconcerted when he heard of the proposed Press visit, and phoned me to say that he felt the advance publicity was premature, and that with Naismith he had organised a watch at longer wavelengths at the Radio Research Station of DSIR. Would it not be wiser, he suggested, to await the outcome of the observations? I reminded him that our previous radar observations had by now been thoroughly expounded in a scientific journal, so we finally compromised by preparing a joint account of the history of earlier investigations at both establishments which we distributed to the Press during their visit to AORG. Appleton and Naismith had in fact recognised short-scatter echoes as a distinct type as early as 1932 and believed they might arise from meteoric ionisation. What had been lacking until 1945 was any convincing proof that all, or even most, of the transient echoes were produced in this way. One cannot but feel admiration for the magnificent spurt made by Appleton when he realised that my team, hotly pursued by Lovell's group, were at the winning post of this particular scientific race.

The Giacobinid shower of 1946 proved to be a tremendous success. It arrived in the early hours of 10th October, and although as the visual watchers reported, the activity was not so intense as had been predicted, it was rich enough to produce a great surge of echoes on the radar screens. Lovell recorded a peak rate of about 200 echoes per minute. Papers were now published by all of us, both on the Giacobinid shower and on radar investigation of meteors. It was evident from the Appleton-Naismith papers that they had been handicapped by observing at a longer wavelength, about 12 m, where the ambient ionisation of the E layer was a complicating influence. Lovell and I had the advantage in operating at about 5 metres wavelength, which effectively filtered the intense ionisation trails, unquestionably associated with incoming meteors. We had been fortunate in choosing equipment which distilled the essence of the phenomenon.

The Giacobinid shower was additionally memorable to Parsons, Stewart and myself at AORG because, after improving our cathode ray tube display, we succeeded in recording for the first time the faint fast-moving echo from the ionisation in the immediate vicinity of the approaching meteor. Previously we had seen only the main echo at right angles to the trail. Now we were able to distinguish the "head" echoes associated with the larger, fast meteors, and measure their speeds ($\sim$ 23 km per second for the Giacobinids.)

In the following years Lovell's group at Jodrell Bank intensively pursued radar astronomy in a most comprehensive manner, studying meteor activity in great detail, and introducing new techniques. In 1948 J. A. Clegg devised a simple and valuable method of determining the radiants of meteor showers. He demonstrated that by looking due East at low elevation, meteors from a given radiant would suddenly become observable, at right angles to the radar beam, at very long range. The sudden appearance of a succession of echoes, followed by a characteristic decline of range allowed the shower to be identified. By rotating the aerial to a more southerly azimuth, the characteristic onset of the shower could be again observed. By a consideration of the geometry, Clegg was able to ascertain the directions of the meteor showers from a single radar.

About the same time, a new method of velocity measurement was demonstrated at Jodrell Bank by C. D. Ellyett and J. G. Davies using an ingenious method originally suggested by N. Herlofson. As a meteor passes at its nearest distance to the radar, at right angles to the aerial beam, a Fresnel diffraction pattern is produced in the reflected radio waves in just the same manner as in optics. Oscillations of amplitude occur as the echo from the newly-forming second half of the trail adds and subtracts alternately in and out of phase with that from the first half of the trail. Knowing the distance of the meteor trail, and the rapidity of the Fresnel oscillations, the velocity of the meteor could easily be calculated.

With the aid of these ingenious and subtle techniques the Jodrell Bank group made great strides in meteor astronomy. At the same time important contributions were being made in Canada under the joint direction of D. R. W. McKinley of the National Research Council (NRC) and P. M. Millman of the Dominion Observatory. Most of their work was at longer wavelengths, about 8 m, using a variety of radar methods and combined visual watches. I shall return to the subject of meteor astronomy later to describe certain developments since 1950.

THE CONCEPT OF THE 250 FT RADIO TELESCOPE

In the pursuance of the original quest for cosmic ray echoes Lovell and Clegg in 1947 laboriously constructed a fixed, upward-looking parabolic reflector 218 ft in diameter with a reflection surface consisting of miles of wire like a spider's web held on to a frame near the ground. By the autumn of 1947, however, after a few fruitless experiments, Lovell decided to abandon the cosmic ray project. For by now he was captivated by the great vista of research opened up by discoveries in radio astronomy. The 218 ft reflector, with its large collecting area, would clearly provide a useful transit telescope for certain investigations. But valuable as the instrument might be, it was irksome to be tied to a vertically-directed beam which could only be slightly tilted by displacing the focus. Lovell soon had visions of a giant steerable reflector which could explore the radio phenomena of the solar system, the Galaxy and the universe. He imagined a super telescope which could be regarded as the radio counterpart of the great optical telescopes like the 200 inch Hale telescope at Mt. Palomar. Of course, the radio telescope could not be comparable in angular discrimination, since the beamwidth and hence the resolution is approximately λ/D radians, where λ is the wavelength and D the diameter of the aperture, for radio wavelengths are about a million times longer than optical. Nevertheless, a huge steerable reflector, say 250 ft in diameter, would be a great instrumental advance in the radio exploration of space.

It was fortunate for Lovell that Blackett was Professor of Physics at the University, for his reaction to Lovell's proposal was immediately favourable. Lovell could not have had a better protagonist in an influential scientist of Blackett's reputation.[1] I well remember during my visits to Manchester in 1945–6 hearing Blackett propound the opinion that research groups with large and powerful instruments would lead world progress in contemporary physics. Several other scientists in the department held more conservative views, and some questioned whether the introduction of radar systems in the research

[1] Professor P. M. S. Blackett, FRS, Nobel Laureate, subsequently Lord Blackett. He left Manchester in 1953 to become Professor of Physics at Imperial College, London, and was elected President of the Royal Society in 1965.

projects could be even properly regarded as physics. Something of the same conventional attitude probably influenced astronomers at the time of Jansky's and Reber's discoveries, namely, that radio methods were outside the traditional province of astronomy. How wrong the reluctance to introduce engineering techniques to serve research in physics has proved to be! Blackett, however, held no doubts, and his part in the lobbying of notable scientists and institutions was a key factor in the successful backing of the proposal to build the giant radio telescope at Jodrell Bank. The important role of the research sponsor is often too easily forgotten when great facilities are provided, and credit is showered on subsequent generations of research students who are the beneficiaries of the initial vision and enterprise. Blackett enlisted the support of the Council of the Royal Astronomical Society which set up a special committee with Sir Edward Appleton as chairman, and a membership which included the Astronomers Royal for England and Scotland, Ratcliffe and Ryle from Cambridge, and myself, and we all gave the venture our blessing. The story of the construction of the Jodrell Bank 250 ft radio telescope after the survival of many vicissitudes, I shall tell in a later chapter. For I must now resume my account of the development of radio astronomy at other research centres.

RADIO ASTRONOMY COMMENCES AT CAMBRIDGE

After Martin Ryle had graduated in physics at Oxford in 1939 he intended to join Ratcliffe's group at Cambridge to pursue radio research on the ionosphere. But the war intervened, and Ryle found himself drafted to TRE to work on the development of airborne radar. Ratcliffe was also at TRE as a divisional leader, except for a period in 1940 when he organised the Army Radio School at Petersham, where I was fortunate enough to be taught by him, for he had a remarkable capacity as a teacher rapidly conveying both enthusiasm and a clear understanding of radio physics. Ratcliffe kept in touch with Ryle during the wartime years at TRE, and when Ratcliffe returned to Cambridge in 1945 to re-plan his department at the Cavendish Laboratory he invited Ryle to join the research staff. By now, however, Ryle's enthusiasm for ionospheric research had largely evaporated, and he felt the urge to branch out into some new field. Ratcliffe then suggested to Ryle that an alternative to ionospheric research would be to investigate the intense solar radio noise associated with sunspot activity that I had reported during the war years. As the sunspot cycle was at the increasing phase, only a few years from the maximum, it seemed likely that there would be excellent opportunities for radio observations of sunspots and flares. With sufficiently sensitive equipment it might be possible to detect abnormal radio emission even during minor solar disturbances. Ryle readily agreed to the proposition, and so a great era of radio astronomy research at Cambridge commenced.

Ratcliffe stressed one essential requirement as a basic necessity for progress in research, that new techniques of observation should be evolved. Although Ratcliffe did not participate directly in Ryle's radio astronomy research even at this early stage, the significance of Ratcliffe's encouragement should not be underestimated because it secured the foundations and conditions for Ryle's initial programme. Unfortunately, there was scarcely any money then available at the Cavendish for new equipment, so Ryle was soon scrounging and foraging amongst the establishments and military workshops, with the happy connivance of the Services and Ministries, for a great assortment of radio and radar equipment. Other ex-members of TRE joined the team, amongst them Graham Smith and A. Hewish who were later to play a particularly valuable part in subsequent advances in radio astronomy.

At the same time that Ryle was making preparations for solar radio research in England, another group had embarked on similar work in Australia. From the start there has been a considerable degree of parallelism between the aims of British and Australian radio astronomers, inevitably leading at times to keen rivalry. I shall be comparing and contrasting methods and achievements in the two countries which on occasions have seemed like an England versus Australia test match. Nevertheless, the outcome of the comparative methods and results of these radio observatories and the friendly competition between them has been extremely beneficial in two respects. The continual checking and re-checking of results has been essential where scientific techniques have been strained to the limits, in order to ensure the elimination of preliminary uncertainties. Further, observations from the two hemispheres have been complementary in completing sky coverage, a necessity in providing a comprehensive view of the universe.

THE START OF RADIO ASTRONOMY IN AUSTRALIA

The Australian Radiophysics Laboratory was originally formed in 1939, with a staff of physicists and engineers, to undertake the development of defence radar in the Pacific theatre of military activity. The Laboratory was organised as part of the Australian Commonwealth Scientific and Industrial Research Organisation, CSIRO. At the end of the war it was decided that the Radiophysics Laboratory should continue within CSIRO as a normal Division, henceforth to be devoted to radio research free from any military commitments, and in 1945 E. G. Bowen was appointed Chief of the reconstructed Radiophysics Division of CSIRO. Bowen had already had an interesting career in radio and radar research. After graduating in physics at Swansea, he carried out research on radio atmospherics under Appleton at King's College, London. In 1935 he was invited to take a post in Watson-Watt's radio group at DSIR, and there was some mystery about the research he was required to do until it transpired, under the cloak of secrecy, that he was to become a member of Watson-Watt's pioneer team in the origination of radar. In the following year Bowen and his colleagues were transferred to the Air Ministry which was then entrusted with the responsibility for organising the development of this vital technique. In the early years of the war Bowen was attached to Sir Henry Tizard's staff, and after taking part in the establishment of the Radiation Laboratory of the Massachusetts Institute of Technology (MIT) he was drafted to Australia in 1944 to advise on radar in the Pacific zone. When Bowen took up his appointment at CSIRO in 1945, the senior member of his staff was J. L. Pawsey, an Australian physicist, who had carried out post graduate radio research with Ratcliffe at Cambridge, and had worked during the war at the Australian Radiophysics Laboratory. Amongst the new recruits to the staff who were to figure prominently in the future progress of the Division were J. G. Bolton and J. P. Wild whose careers had independently followed curiously parallel paths. Although unacquainted with each other both were born in Sheffield, England, read Physics at Cambridge, served in the Navy in the Pacific zone as radar officers, and after the war applied for posts at CSIRO in response to the same advertisement.

The first task confronting Bowen and Pawsey was to decide the research programme. Many subjects were considered and eventually filtered down to two major items; rain physics (including artificial rain making—an important problem for Australia), and solar radio astronomy. They knew of my paper on intense solar emission and of Southworth's measurements of radiation from the quiet Sun. They had also heard of unpublished reports of solar radio noise being noticed by radar stations in New Zealand in early 1945. It was now agreed that while Bowen would personally direct the meteorological radio research at CSIRO, Pawsey would be responsible for further investigations of solar radio noise. As Pawsey had been the driving force in developing an all Australian radar system for the defence of the north of the continent, he was now able to recruit the assistance of the Australian Air Force radar stations and personnel in making a series of observations of solar radio noise from October 1945 to February 1946. The data provided the basis for a paper by Pawsey's team in which they concluded that at a wavelength of 1·5 m a rough correlation existed between the level of solar radio noise and the area of sunspots on the Sun's disk.

INTERFEROMETERS AND THE SUNSPOT SOURCE

The problem that now confronted Pawsey in Australia and Ryle in England was how to determine the size of the radio source in order to confirm the obvious inference that the radio emission originates from the sunspot region. Independently, both Pawsey and Ryle answered the questions with the aid of interferometers. Strictly in order of time, Pawsey's team (McCready, Pawsey and Payne-Scott, 1947) obtained results first, but published later than Ryle (Ryle and Vonberg, 1946), which shows how confusing it can be in attempting to establish precedence. We may fairly regard them as independent, simultaneous investigations.

In 1946, a typical radar aerial at a wavelength of about 1·5 m had a beamwidth of 10° or more, quite useless for locating a small region on the Sun, itself only half a degree in diameter. Pawsey argued that if radar stations situated on a coastal cliff looking out over the sea watched the Sun rising, then the crests and troughs of radio waves received directly and those reflected from the sea would mutually interfere with each other, and split the beam into a series of lobes each sufficiently narrow to allow the size and position of the source to be located. The method is precisely analogous to Lloyd's mirror experiment of 1837 in which he demonstrated optical interference fringes by reflecting light waves from a mirror at glancing incidence. Two radar sites, Dover Heights at 280 ft above sea level, and Collaroy at 400 ft, proved ideally situated for the radio experiment, and the great sunspot of February 1946 and its subsequent reappearance in later months presented an excellent chance to test the method. The ratio of maximum to minimum intensities clearly indicated that the radio noise came from the vicinity of the sunspot groups and that the source was of similar size.

Ryle's interferometer was an elegant analogy of the Michelson optical interferometer designed to measure the optical diameters of stars. Ryle connected together two horizontally separated aerials, the combined signal producing interference maxima and minima as the direction of a source moves across the sky. One advantage of Ryle's method was that the width of the interference lobes could be altered at will by changing the spacing between the aerials. Also, by looking at high elevation, refraction in the atmosphere and ionosphere, evident in Pawsey's observations at low elevation, could be avoided. Ryle and Vonberg also employed a version of the Dicke comparison technique in recording the output, and in July 1946 they successfully demonstrated that the size of the sunspot radio source was comparable with the visible region.

Subsequent developments of many types of high-resolution radio telescope systems have been profoundly influenced by interferometry. As Pawsey and his co-workers pointed out in their 1947 paper, a simple two-element interferometer furnishes one Fourier component of a source distribution, and a complete picture of a source can be calculated from measurements at many spacings. The ultimate resolution is then determined by the maximum spacing. Pawsey realised that changing the height of a cliff edge interferometer was hardly feasible and noted that "a different interference method may be more practicable". Without doubt he had Ryle's horizontally spaced interferometer in mind because it possesses just the required adaptability and versatility. In subsequent years Ryle's group progressively exploited the basic principles of interferometry and Fourier analysis; and Ryle's name, more than any other, has been familiarly associated with interferometry and aperture synthesis in radio astronomy. The Australians and other groups have also shown remarkable ingenuity in the manipulation of aerial elements in various forms of interferometry, and in combinations of many elements together to produce multiple beams or single beams. I shall discuss later, in Chapter 4, the various types of radio telescopes, as well as the merits of the very large steerable reflectors typified by the Jodrell Bank 250 ft parabolic dish.

EARLY STUDIES OF THE SOLAR ATMOSPHERE

Despite the notorious complexity of solar phenomena, a pattern of behaviour of solar radio emissions was beginning to emerge. A radio noise storm was now known to accompany certain sunspots with an outwardly directed beam of circularly polarised radiation. Even more intense were the sudden outbursts of radiation often accompanying bright solar flares, which occur spasmodically in the vicinity of active sunspots. Major solar flares near the centre of the solar disk are frequently followed in a day or so by magnetic storms at the Earth, indicating that a stream of charged particles, ions and electrons, must have been ejected from the Sun at the time of the flare taking a day or so to reach the Earth. Could these particles emerging from the Sun's atmosphere somehow originate the radio outburst? Striking support for this view came from the time delays of several seconds between bursts at different frequencies recorded by Payne-Scott, Yabsley and Bolton (1947) who found that outbursts occur later at lower frequencies. An explanation in terms of a stream of particles moving outward through the solar atmosphere was readily forthcoming. An ionised gas has a natural frequency of oscillation, the critical or plasma frequency, at which the gas is opaque and reflects or absorbs radio waves. Higher in the solar ionised atmosphere where the density becomes less, the plasma frequency is lower. If the corpuscular stream somehow excited emission at the plasma frequency as it moved out from the Sun, the observed delays would be accounted for.

The intense sunspot and flare emission were exceptionally startling phenomena, but it was characteristic of Pawsey that he should be seeking the answer to a simple basic question, namely, what really was the base level of radio output from the Sun in its quiescent state? Daily recordings enabled him to assess the residual component in the absence of enhanced emission. The lowest level could be expected to represent thermal radio emission from the Sun. Curiously enough, the radio brightness temperature turned out to be about a million degrees at 1·5 m wavelength. During a visit to CSIRO, D. F. Martyn, well-known for his theoretical ionospheric research in Australia, was shown Pawsey's analysis. Martyn pounced on it with the alacrity of a theorist presented with a key experimental result. He at once realised the explanation, the quiet Sun radiation was emanating from the solar corona where the electron density would be sufficient to render it opaque at metre wavelengths. Previous evidence of extremely high temperatures in the solar corona, about a million degrees, had been derived from optical spectral lines, although the heating process was a matter for conjecture. At shorter wavelengths, the solar atmosphere would be opaque closer to the surface where the temperature would be less. Martyn saw the whole picture, the distribution of radio brightness across the face of the Sun and the dependence on wavelength. Martyn was so eager to publish his calculations that his preliminary article might well have appeared in advance of the experimental results which prompted the work, had not Pawsey managed to negotiate a brief account in the same issue of *Nature* (2 November, 1946). Actually, unknown to the Australians, a theoretical assessment, with similar conclusions, had already been made independently by V. L. Ginzburg in Russia earlier in the same year. Following a suggestion by

N. D. Papalexi in 1945, that radar echoes from the Sun might be practicable, Ginzburg examined theoretically what reflection coefficient might be expected. He concluded in 1946 that absorption would be a serious attenuating factor in attempting to obtain radar echoes, and instead he suggested that it would be easier to observe radio emission from the solar atmosphere, and that the radio brightness temperature would increase with wavelength and attain the coronal temperature of a million degrees. However Martyn's papers (1946, 1948) provided the first detailed treatment of the brightness distribution of the quiet Sun at different wavelengths, which he derived from the basic equations of radio propagation. Almost identical results were obtained soon afterwards via a quantum mechanical treatment by the German astronomer Unsöld and by the Swiss astronomers, Waldmeier and Müller. Although later work refined the calculations it was clear that the essential points governing thermal radio emission from hot ionised gas such as the solar atmosphere, or the gaseous nebulae surrounding hot stars, were now well understood, and that radio observations could be especially helpful in deducing electron densities and temperatures. In these instances, the radio emission was simply an extension of classical concepts of thermal radiation.

THE DISCOVERY OF THE RADIO SOURCE, CYGNUS A

The most intriguing radio phenomena were the abnormally intense radiations such as those associated with solar activity and the unaccountably strong radiation from the Galaxy at long wavelengths. In 1946, my group made yet another startling discovery, a powerful discrete radio source in the sky. At that time a large proportion of the effort of the team, comprising Parsons, Phillips, Stewart, and myself as leader, was devoted to radio astronomy, and we shared the investigations of solar and cosmic noise and radar echoes from meteor trails. Stewart specialised more particularly on the meteor research, and Phillips on the cosmic radio noise which we knew from the work of Jansky and Reber was principally produced by the Galaxy. To map the distribution of the cosmic noise, an aerial system consisting of 4 Yagi arrays directed horizontally and rotatable in azimuth was erected on a modified radar receiver in Richmond Park as shown in Plate 2.1. The gain of the aerial

Plate 2.1 The equipment in Richmond Park used for the first observations of Cygnus A in 1946

system was increased by reflection from a wire-netting mat covering the ground to give a resultant beamwidth of 12°. While Phillips was engaged in recording the received noise he noticed that in one direction in the sky the noise level often varied irregularly, usually with a period of the order of a few seconds. The variable radiation appeared to be coming from a direction in the constellation of Cygnus. We could locate the position of the source only to within about 2°, and there was no obvious optical clue to its identification with visible stars. Nevertheless, we concluded that only discrete sources could produce such fluctuations (*see* Figure 2.1). As we were fully cognisant of the large changes and surges of radio emission from the Sun during sunspot and flare activity, we believed that the Cygnus source must be a similar but more distant source of powerful localised disturbances. We published our results (Hey, Phillips, and Parsons 1946, 1948) first in *Nature* and then in more detail in the Proceedings of the Royal Society.

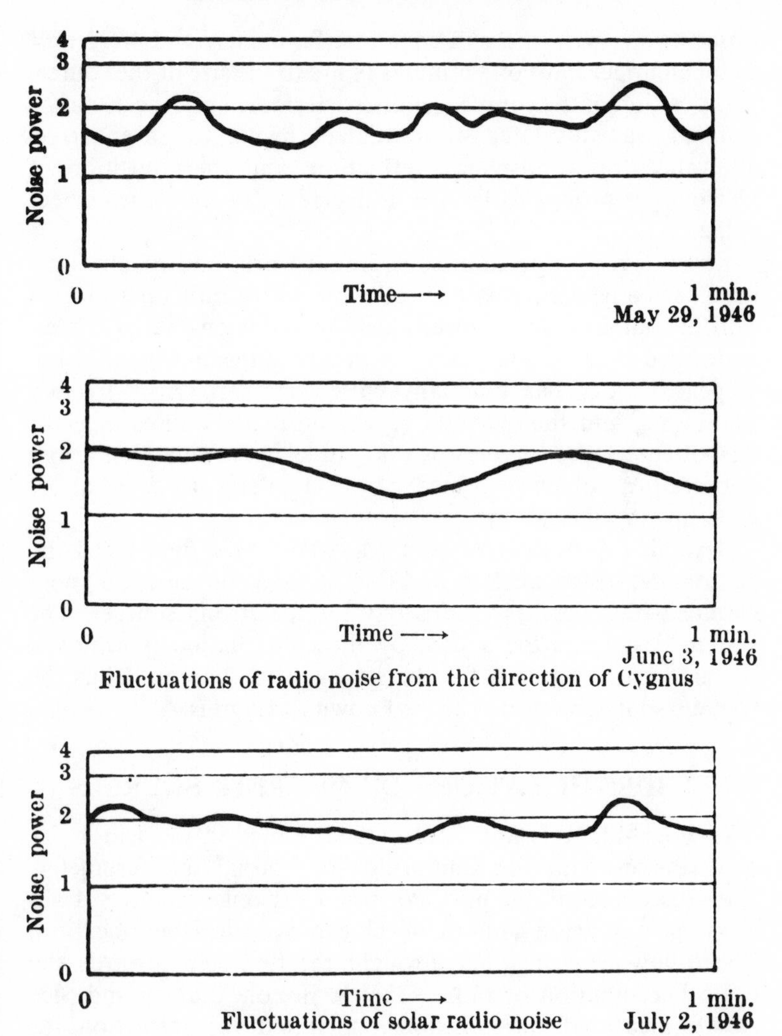

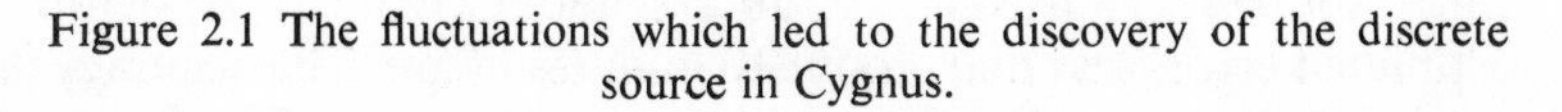

Figure 2.1 The fluctuations which led to the discovery of the discrete source in Cygnus.

At once the attention of radio astronomers elsewhere was diverted to this amazing discovery. As it happened, the interferometers that had been set up for solar observations were ideally suited for detecting and locating discrete sources. In Australia, J. G. Bolton and G. J. Stanley began a series of observations with the cliff-edge interferometer, and were able to set an upper limit of 8 minutes of arc to the size of the source. The position of the source was also determined but the accuracy claimed was not later substantiated, owing to an incorrect allowance for atmospheric refraction, and their position in Declination proved to be over a degree away from the correct value.

Errors were prone to arise in initial work, and reappraisals have been a necessary part of the evolution of radio astronomy. Throughout the years reported positions and intensities have been bedevilled by excessive claims of accuracy made in good faith. Repeated work has gradually chiselled down errors to very fine limits, and the eventual revelation of the causes of error has often proved of inestimable value in leading to a close appreciation of unsuspected physical effects as well as of experimental limitations.

Several other discrete sources were now discovered by Bolton and his co-workers and they chose a convenient nomenclature which persists to this day for the strong sources. The system designates the source by the constellation in which it is found and a letter A for the strongest, and so on. Thus the Cygnus source continues to be known as Cygnus A.

FIRST IDENTIFICATIONS OF DISCRETE SOURCES

In England, Ryle and Smith (1948) joined in the search for new sources with the Cambridge two aerial interferometer. They discovered in the northern sky the strongest source of all, Cassiopeia A, again without any clue to its optical identification. The following year, 1949, brought the first step towards the optical recognition of radio sources. Bolton, Stanley and Slee (1949) measured the positions, corrected for refraction, of three sources, Taurus A, Virgo A, and Centaurus A, with estimated errors less than about 10 minutes of arc. The accuracy was sufficient for them to suggest the astronomical objects associated with all three sources. Their tentative identifications have all subsequently been confirmed by more accurate location of the radio positions. The sources are as follows:

Taurus A: The Crab Nebula, the expanding shell of a supernova in the Galaxy

Virgo A: The extragalactic nebula NGC 4486 (M87)
Centaurus A: The extragalactic nebula NGC 5128.

All three are exceptional astronomical objects. The Crab Nebula is the best known example of a supernova, or more precisely, the remnants of a supernova explosion recorded by the Chinese in 1054. Its comparative proximity to the Earth (about 3500 light years away) and clear visibility have made it the most frequently studied object in the Galaxy. Virgo A is an extraordinary external galaxy with a remarkable jet extending from the nucleus. Centaurus A is a peculiar galaxy, visible in the southern hemisphere with a dark dustband straddling across it.

FURTHER OBSERVATIONS OF THE SUN AND MOON

Radio studies of the Sun and Moon naturally continued to receive attention. Much valuable research could be done, even with radio beams large compared with the $\frac{1}{2}°$ angle subtended by the Sun or Moon, simply by measuring the total radiation. For instance at CSIRO, Piddington and Minnett (1949) measured the radiation from the Moon over a lunar cycle at a wavelength of 1·25 cm. They found that the temperature indicated by radio lagged behind the visible phase of the Moon because waves penetrate slightly through the solid material, so the radio temperature is partly that of the sub-surface which heats more gradually and cools more slowly than the surface. They interpreted their results as indicating a thin layer of dust over porous rock or gravel at the Moon's surface.

Solar radio observatories became established and particularly notable has been the centimetric watch on the Sun that has been maintained since 1946 by Covington at NRC, Canada. At a wavelength of about 10 cm he found a close correlation between radio emission and sunspot area.

To obtain a radio picture of the Sun presented a difficult problem, and radio astronomers were quick to avail themselves of the indirect aid in resolution offered by nature in the way of solar eclipses, which provide opportunities for distinguishing radiation from different parts of the Sun. For example, Christiansen, Yabsley and Mills observed the partial eclipse of 1 November 1948 from three separate sites in Australia. The eclipsing of each active sunspot region was accompanied by a sudden fall of intensity. The intersections of the Moon's rim as seen from the different sites demarcated the location and size of the active regions.

It is harder to decipher the radio emission across the quiet Sun even with a total eclipse because the pattern of the distribution over the whole solar disk does not lend itself to easy analysis. The first total eclipse to be observed by radio was on 20 May, 1947 when the path of totality crossed the South Atlantic ocean and both the Americans and Russians organised seaborne expeditions. The American observers, led by Hagen of the Naval Research Laboratory USA, watched the eclipse at $\lambda = 3{\cdot}2$ cm, while the Russian group under Khaikin and Chikhachev observed at $\lambda = 1{\cdot}5$ m. Solar activity hampered attempts to deduce the quiet Sun distribution, but Hagen concluded there was radio brightening at the solar rim at 3·2 cm wavelength, and the Russians at 1·5 m found that 40% of the radio Sun remained exposed at optical totality, both conclusions being in accord with Martyn's theory. But interest in radio observations dwindled in later years, for the results scarcely repaid the effort involved, so eclipse expeditions were largely superseded by the development of high-resolution aerial systems enabling the Sun to be examined at any time.

LOOKING AHEAD

By 1949, it was abundantly evident that radio astronomy had uncovered a rich field for further research, with many unsolved problems and the challenge presented by the need to improve techniques of observation. It was now obvious that radio methods were destined to have a major impact on astronomy. The achievements so far had been sufficient to warrant the expansion of existing research centres and the inauguration of new ones. The original teams, in some instances with an initial staff of three or four, were being granted increased resources in manpower and equipment. By 1949 there were radio observatories, or the nuclei of new ones, in many countries, for instance, England, Australia, Holland, France, Russia, Canada, and USA. For my own part, unfortunately, the story was different. The happy era of radio astronomy at AORG was over. In foreign affairs, international tensions seemed to be mounting again and new tasks confronted the group. I was appointed head of AORG and consequently for several years, I had to relinquish my research, although not my interest in radio astronomy. In these years, certain fundamentally important conclusions were emerging which greatly influenced the future course of radio astronomy, as I shall describe in the next chapter.

CHAPTER 3

Two Crucial Years, 1950–1

THE APPARENT FLUCTUATIONS OF CYGNUS A

In retrospect, it seems strange that four years had to elapse before it was realised that the fluctuations found in Cygnus A were not intrinsic variations of the source after all. One may ask, why was the straightforward test not made earlier to find out whether identical scintillations were recorded simultaneously at widely separated sites? The apparent twinkling of a star or of a distant street lamp due to atmospheric irregularities along the light path is common knowledge. Might it not be that the Earth's atmosphere or ionosphere could impose scintillations on radio waves? Two factors misled radio astronomers, causing them to ignore the idea; firstly—a rather weak excuse one must admit—according to prevailing concepts of the Earth's upper atmosphere there seemed no obvious reason why metre wavelengths should be so affected; secondly, a slightly better reason, although Cassiopeia A also showed fluctuations at times they were generally less severe, and so, it was argued, variations must be an inherent property of the source.

Suspicions were first aroused when Bolton and his co-workers, comparing records of Cygnus at cliff edge sites in Australia and New Zealand, failed to find any correlation in the variations at the two sites. The results were communicated privately to British radio astronomers at Cambridge and Jodrell Bank who then organised a series of tests with different separations between recording sites. The 1950 publications by Smith at Cambridge, and by Little and Lovell at Manchester, demonstrated that simultaneous correlation was good at sites within a few km of each other, but not for the separation of 210 km between Jodrell Bank and Cambridge. These results, together with subsequent studies, led to the conclusion that the variations of intensity were imposed on the radio waves as they passed through the ionosphere, and were caused by diffraction in the F region from irregular clouds of ionisation of about 5 km average size. Actually it was found that some well-separated sites could show a correlation between the fluctuations, but with a time delay, as the ionospheric irregularities were blown across by winds in the upper atmosphere. Only very slight variations of electron density were sufficient to distort the wavefront, hence producing large amplitude scintillations. Paradoxically then, the fluctuations that led to the first detection of Cygnus A were not inherent in the source after all. Nevertheless, the argument that they originated from a localised source remained valid. Fluctuations tend to average out in large sources, just as the Moon does not appear to twinkle. Another influential factor is the elevation of the source because the path length through the irregularities is longer at low elevations. Mainly for this reason Cassiopeia A, usually seen at high elevations, is less susceptible to fluctuations. Over the years, one research worker, more than any other, has persistently investigated various forms of scintillation and their interpretation; as we shall see later, the studies by Hewish at Cambridge have yielded an exceptional harvest of interesting results.

SYNCHROTRON RADIATION

In the same year, 1950, a most promising hypothesis, with far reaching impact on astrophysical theory, was proposed to account for the intense radio emission, namely that the radiation might arise from electrons moving with relativistic speeds in weak magnetic fields. The theoretical principles governing such radiation had been expounded long ago by Schott[1] in 1912. His work did not evoke much interest at the time except as an academic exercise. The theory attracted renewed attention in the late 1940s when devices were constructed to accelerate electrons to relativistic speeds for experiments on the bombardment of atomic nuclei by high speed particles. In one such machine, the synchrotron designed by the General Electric Company, USA, where electrons were accelerated to extremely high speeds in strong magnetic fields, an unexpected radiation of light occurred. With the revival of interest, several papers

[1] G. A. Schott, Electromagnetic Radiation, Cambridge Univ. Press, 1912.

appeared in Russia (Artisimovich and Pomeranchuk, 1946, Ivanenko and Sokolov, 1948) and in USA Schwinger, in 1949, published a paper entitled "On the classical radiation from accelerated electrons" presenting a detailed analysis of the problem previously treated by Schott. Theory shows that the radiation from electrons moving with relativistic speeds in magnetic fields would cover a very wide band, and the spectral maximum would depend on the electron energy and magnetic field strength. Hence for certain fields and speeds, radio emission would predominate, for higher values the radiation would be optical. Radiation generated in this manner became known as synchrotron emission since it was first observed in the synchrotron machine. A magnetic field of 10^{-4} or 10^{-5} gauss, and electron velocities of 0·9999999 times the velocity of light (energy about 1000 MeV) would be appropriate values for generation of synchrotron radio emission in astronomical sources. A special feature is that large amounts of power can be radiated by comparatively few electrons in weak magnetic fields because the electrons possess such high energies by virtue of their velocity. High energy protons were known to exist in space in the form of cosmic rays and it has always been assumed that they must be accompanied by high energy electrons. In 1950, the physicists Alfvén and Herlofson in Sweden were the first to postulate that the synchrotron mechanism could be responsible for the radio emission from the intense discrete sources. In the same year, quite independently, Kiepenheuer in Germany suggested that the non-thermal radio emission from the Galaxy may be produced in a similar manner. More detailed developments of the theory followed in a series of papers published in Russia by Ginzburg and Getmansev. Although the synchrotron mechanism is not the only process of abnormal radio emission, it is undoubtedly the most prevalent cause of intense radio power from astronomical sources.

DETECTION OF THE 21 cm HYDROGEN LINE

The prediction of the 21 cm line radiation from neutral atomic hydrogen and its subsequent detection in 1951 was an achievement in the elegant classical manner one always imagines the course of research ought to follow. Starting from Oort's perception in 1944 of the potential value of a radio spectral line as an indicator of the dynamics of the Galaxy, van de Hulst had predicted that a line at 21 cm wavelength from interstellar hydrogen should be observable. The next stage was to design and construct radio spectrometers, special types of sensitive receivers for spectral line observations. The temperature of interstellar hydrogen was expected to be no more than about 100 K, and generally the effective brightness temperature would be much less owing to the low concentration of interstellar gas and to the dispersion in Doppler frequency shifts arising from varied motion of the gas. Obviously then, a specialised type of receiver had to be designed to measure complex line profiles of weak intensity. In fact, it was six years after van de Hulst's prediction before appropriate equipment had been developed. The Dutch group would almost certainly have been first to detect the line but for an unfortunate accident, a conflagration which destroyed their original radio receiver. As it turned out, Ewen and Purcell of Harvard observed the line radiation six weeks before Muller and Oort in the Netherlands, followed soon afterwards by Christiansen and Hindman in Australia. Curiously enough, Ewen and Purcell's first detection of the line was made during a visit by van de Hulst to Harvard. All three groups employed different versions of the same method; a narrow band within the anticipated profile was compared with a reference signal by means of the Dicke switching technique, while the line reception band was swept in frequency to cover the profile.

DETECTION OF THE ANDROMEDA NEBULA

Since our Galaxy generates radio waves it was to be expected that other galaxies like our own would radiate in a similar manner. At Cambridge, Ryle, Smith and Elsmore (1950) noticed that four faint sources appeared to correspond to nearby normal galaxies; one of them was Andromeda, the nearest spiral galaxy, its visible nebula subtending about 3° by 1°. Hanbury Brown and Hazard, at Manchester, examining the radio source in 1951 with the 2° beamwidth of the 218 ft fixed parabolic reflector, found that the radio source was larger than the visible spiral, and so were able to plot the first radio map of an external galaxy. The intensity of the radio emission was comparable in strength with that generated by our own Galaxy.

IDENTIFICATION OF CYGNUS A AND CASSIOPEIA A

At this time, the attempts to find and identify more "radio stars" the name then applied to discrete sources, had become one of the main preoccupations of radio astronomers. The paucity of identifications with visual objects was baffling. Ryle, Smith and Elsmore had located about 50 "radio stars" and concluded that most of them must represent "a hitherto unobserved type of stellar body". The sources were obviously immensely strong radio emitters in comparison with their visibility since attempts at optical recognition had been so largely defeated. Evidently, if the enigma was to be resolved, the radio position would have to be determined with great accuracy, followed by a very careful scrutiny in the specified directions with very powerful telescopes. So Graham Smith at Cambridge embarked on a rigorous study of the precision of interferometer methods, and in 1951 he refined the positional measurements of several sources including the two strongest, Cassiopeia A and Cygnus A. I well recall the excellence of Graham Smith's work, for I was the external examiner for his research thesis, for what must have been, I think, the first Ph.D. degree award in radio astronomy at Cambridge. He derived the following positions:

	Right Ascension			Declination
	h	m	s	
Cassiopeia A	23	21	$12 \cdot 0 \pm 1$	$58° \; 32' \cdot 1 \pm 0' \cdot 7$
Cygnus A	19	57	$45 \cdot 3 \pm 1$	$40° \; 35' \cdot 0 \pm 1'$

These accurate locations led to exciting identifications of great significance. Smith communicated the positions to Baade and Minkowski at the Palomar Observatory so that the 200 inch telescope could be brought to bear to examine the field of view in these directions in detail. Meanwhile a prelude to the identifications was being accomplished elsewhere. Dewhirst, of the Cambridge astronomical observatory, pointed out that a curious faint nebulosity appeared to coincide with the position of

Cassiopeia A. And in Australia, Mills and Thomas had measured the position of Cygnus A and postulated that a faint extra-galactic nebula, 7 seconds away in Right Ascension, might be the radio source. Smith's more accurate position of Cygnus A was only 1 second away from the nebula. These tentative identifications were subsequently confirmed in the photographs obtained by Baade and Minkowski with the 200 inch Palomar telescope which fully revealed their unusual character. So the two strongest discrete sources in the sky, Cassiopeia A and Cygnus A were now, after five years, at last identified, and the nebulae associated with them were found to represent two entirely different, contrasting types of astrophysical phenomena.

The nebulosity associated with Cassiopeia A exhibited such fast-moving filaments that changes in detailed appearance were discernible in only a few months time. It was obviously a relatively near source, within our own Galaxy. Later studies established that the nebulous wisps and filaments must be remnants from an unrecorded supernova at about 10 000 light years distance and which, it was estimated, must have exploded about A.D. 1700. The question at once arises; why did the supernova occurring at the beginning of the 18th Century escape observation? There appears to be a twofold explanation: interstellar absorption is high at the centre of the nebula which would correspond to the direction of the exploding star; also, the supernova is believed to be type II, which has lower initial luminosity than a type I explosion such as the Crab supernova.

Cygnus A proved to be a most remarkable type of source, an extremely distant, peculiar double galaxy. The distance of Cygnus A, determined by Hubble's law from the redshift of the spectral lines, was found to be about 550 million light years. The optical faintness could be attributed to the great distance, yet this was the second strongest radio source, showing that it was an amazingly powerful radio emitter. Knowing the distance, the radio luminosity could be calculated, and was found to be about a million times greater than that of a normal galaxy like our own. Consequently, the name "radio galaxy" was henceforth adopted to describe extra-galactic nebulae possessing exceptionally powerful radio luminosity. At this time, no stars other than the Sun had been identified with discrete radio sources and the term "radio star" fell out of use. The identification of Cygnus A marked the beginning of an era in the study of unusual, exceptionally active galaxies. The double appearance of the galaxy associated with Cygnus A, and the high excitation and turbulent velocities indicated by the optical spectral lines, were originally thought to suggest a collision between two galaxies, but the idea eventually had to be discarded. I shall return to a discussion of radio galaxies in Chapter 7.

By the end of 1951 a sure foundation had been laid for the expansion of all branches of radio astronomy. It was that kind of time with everything moving faster. To clarify my account of developments in the next two decades, I must relinguish the attempt to maintain a chronological sequence over the whole field, and treat the different branches separately. Even so, with the multiplicity and complexity of astronomical topics in which radio methods were to play a part, I must run the risk of over-simplification.

It was by now clear that radio astronomy could make a great contribution to astrophysics. The fusion of optical and radio astronomy had already commenced, as exemplified by Oort's interest in the radio structure of the Galaxy, and the cooperation of optical and theoretical astronomers in the identification and study of radio sources. Much sustained observational work lay ahead. It was felt that the days of improvised aerial systems were passing and must give way to the planning of long term projects. If the depths of the universe were to be probed, larger and more sensitive instruments were required. The realisation of large radio telescopes is a story of scientific vision and struggle, competition and controversy, persistence and diplomacy. Before I continue with further research results and discoveries, I will describe some of the radio observatories that were established and the great radio telescopes whose development has depended so much on the character and determination of radio astronomers.

CHAPTER 4

Radio Telescopes and Observatories

THE 250 FT TELESCOPE AT JODRELL BANK

Mention of a radio telescope conjures up in the minds of many people the 250 ft radio telescope at Jodrell Bank. It is, of course, now in the 1970s, yielding place to larger and more accurate instruments. But it must always remain a daring pioneering enterprise in the construction of massive steerable radio telescopes. I have already described how, with Blackett's support, Lovell secured in 1950 the scientific backing for his ambitious proposal. It was believed that the cost would be between £50 000 and £100 000. The Council of the Royal

Plate 4.1 A. C. B. Lovell (*by courtesy of A. C. B. Lovell, Jodrell Bank*)

Astronomical Society fully endorsed the scheme and the impressive list of proposed research projects. Lovell had contacted a talented and enthusiastic engineering consultant, H. C. Husband, who was henceforth to be entrusted with the engineering plans and design. A formal application for £259 000 to build the telescope was despatched to the Department of Scientific and Industrial Research (DSIR) in 1951, and a year of suspense for Lovell followed. Although DSIR were favourably disposed to the proposal, by 1952 the estimated cost had risen to £335 000, an amount which placed far too severe a burden on their capital resources. Consequently an approach for financial assistance was made to the Nuffield Foundation, who generously offered to contribute £200 000 of the required sum. Little did Lovell and the Manchester University authorities realise that clearing this first financial hurdle was merely a preface to gruelling years of struggle and anxiety. Lovell was soon engulfed in a mounting series of obstacles, the foremost being the escalation of costs due to initial underestimates combined with unforseen rises in the price of steel and materials. By 1953 the estimated cost had risen to £460 000; DSIR finally agreed to authorise the additional payment subject to the proviso that they could not be expected to meet any further increases.

The next crisis was precipitated by second thoughts on specification and design. Lovell's original proposal had been made when the most interesting radio phenomena were predominantly apparent at metre wavelengths, and consequently it appeared that a reflector surface of 2 inch wire mesh and a tolerance of ± 5 inches in the parabolic profile would suffice. After the detection in 1951 of the 21 cm hydrogen line, the idea of making a more accurate profile for the central portion of the dish to permit operation at 21 cm wavelength presented a most attractive possibility. The desirability of incorporating such an improvement became compelling, and in order to conform to the new plan, Husband had to alter the design of the supporting frame to improve the rigidity and he decided to make the reflector surface of sheet steel instead of mesh. Meanwhile the escalation of costs continued as construction proceeded, and by 1956 a further £260 000 was needed. An examination instigated by the Treasury led to the publication of painful allegations by the Public Accounts Committee. In a tense situation and apparent impasse, Lovell at one time feared that he might suffer imprisonment as the chief instigator of the deficit that had been incurred. The scientific potentialities of the huge structure now rising in the Cheshire plains were well nigh forgotten, submerged by the financial crisis. It was fortunate that the engineering of the great project had been entrusted to Husband's great skill, and the successful completion of the massive mechanical venture was in sight. Then a remarkable and unexpected event suddenly heralded a change of fortune. In October 1957, the Russian Sputnik was launched and the great radio telescope swung into action to track by radar the carrier rocket and satellite. The achievement of the radio telescope was received with acclamation, and provided a

Plate 4.2 The 250 ft MK I radio telescope at Jodrell Bank (*by courtesy of A. C. B. Lovell, Jodrell Bank*)

welcome boost to national prestige. The Prime Minister, the Rt. Hon. Harold Macmillan, acknowledged the feat in the House of Commons. The struggle to resolve the financial situation continued, but at least there was now a change of heart. Further success followed, for the assistance of the telescope was sought by the Americans in tracking their space vehicles. The Public Accounts Committee decided to retract its earlier report. Finally, with the help of the Nuffield Foundation, the debt was finally cleared. Although Lovell has always maintained a practical interest in space applications, about 99% of the working time of the radio telescope has since been devoted to radio astronomy as originally intended. Research at Jodrell Bank flourished in many branches of the subject, radar investigations of meteors, aurorae, the Moon and planets, radio emissions from the Sun, the Galaxy, and discrete radio sources; and in 1961 Lovell was knighted for his contributions to the progress of science.

INTERFEROMETRY AT CAMBRIDGE

Almost all the developments of radio telescope systems at Cambridge could be described as the gradual evolution of interferometric methods for the study of radio sources. The purposeful and progressive improvements of interferometric techniques to attain higher resolving power and greater sensitivity have continually reaped a rich harvest of results.

The earliest radio interferometer simply consisted of two connected aerials separated along an E–W baseline. Ryle then introduced phase-switching, by alternately reversing the phase of the signal from one aerial, so interchanging the maxima and minima of the interference pattern. A phase-sensitive rectifier at the output measured the difference in the signals for the two positions of the lobes. With this system, background radiation gives no response; only discrete sources smaller than the lobe pattern register a response, thus providing a sensitive method for their detection. The principle of the interferometer is illustrated in Figure 4.1.

After a preliminary survey, known as 1C, of about 50 discrete sources, a large interferometer was constructed consisting of four parabolic cylinder stretched-wire reflectors, situated at the corners of a rectangle, giving an E–W spacing of 580 m and a N–S spacing of 50 m. With this radio telescope at $\lambda = 3{\cdot}7$ m, Ryle and his colleagues in 1955 attempted an extensive survey of radio sources, labelled 2C. Although a considerable

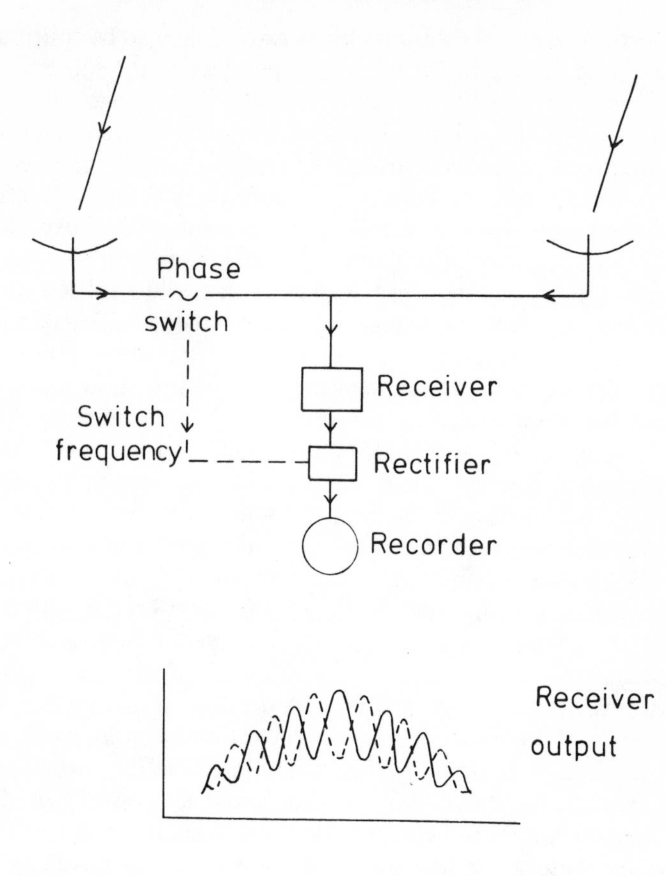

Recorder
output

Figure 4.1 The phase-switching interferometer and its response as a discrete source moves across the aerial pattern.

proportion of the 2000 sources listed proved later to be spurious, the exercise represented a salutory experience in the subsequent attainment of reliable surveys. The 2C survey had suffered from an over-enthusiastic interpretation of apparent recordings, many of which were attributable to the beating together of interferometer records from two or more weak sources, thereby producing illusory sources. Poor agreement with Australian narrow beam observations over a common area of sky limelighted the inconsistencies. It was eventually realised that confusion occurs if the number of sources listed exceeds about 1 source per 25 primary beam areas. With the same interferometer adapted to a shorter wavelength, $\lambda = 1 \cdot 9$ m, in order to reduce the beam area, a new and more reliable Cambridge list of 471 sources, 3C, was published in 1959.

Meanwhile another trend in radio interferometry, variable spacing, was gradually gaining ground. The possibility of deriving source structures in this way had been appreciated by Pawsey and his colleagues in their 1947 paper on solar observations. By measuring the depth of the interference pattern, called the "fringe visibility", over a range of interferometer spacings, source structures can be calculated from the Fourier transform of the fringe visibility expressed as a function of spacing. The theoretical principles have long been known, and were applied to optical interferometry in the latter part of the 19th century. In 1950, Stanier at Cambridge successfully used a two element interferometer with variable spacing to derive the radio brightness distribution of the Sun at 60 cm wavelength. Of course, one direction of baseline only provides a one-dimensional brightness distribution, and observations covering many directions are required to derive complete two-dimensional solutions.

The method of Fourier synthesis can be regarded from another point of view, namely aperture synthesis, as Ryle and Hewish have elaborated in a 1960 paper. The principle can be outlined as follows. Imagine a large aperture divided into small areas as illustrated in the diagram, Figure 4.2. If a radio telescope, the size of a small section, could be placed at each position in turn, and the received signals added up in the correct phase, the result would be equivalent to the whole aperture. Phase can be taken into account by combining

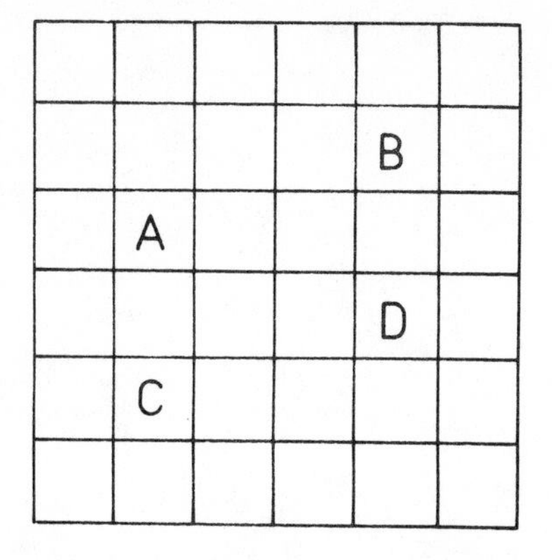

Figure 4.2 Synthesis of an aperture.

positions in pairs, like A and B. Now a pair of aerials taken together form an interferometer. Hence, a variable-spacing interferometer including all relative spacings can synthesise a large aperture. As two elements at C, D would give the same answer as A, B, this simplifies the practical procedure.

Aperture synthesis is an elegant way of achieving the resolution of a very large aperture. The direction of the 'beam' of the aperture can be altered merely by inserting phase differences when computing the resultant. Are there any snags at all in the method? Two points should be borne in mind. The resultant has to be computed after making many separate measurements and is subject to the addition of any errors; furthermore, the statistical addition of receiver noise means that the power sensitivity is less than for reception over the whole aperture.

The two elements of the interferometer need not be equal in size, and the original aperture synthesis systems at Cambridge used a large element at a fixed location in conjunction with a small mobile element. The first synthesis experiment was conducted by Blythe in 1957 using corner-reflector aerials to obtain a detailed map of the Galaxy at $\lambda = 7{\cdot}9$ m, with an effective angular resolution of about 1°.

The need to increase the extent of flat land for placing and moving aerials to simulate very large apertures forced Ryle to seek a new observing site, and with the aid of a generous benefaction from Mullard Ltd and a grant from DSIR, an area comprising

Plate 4.3 M. Ryle (*by courtesy of M. Ryle, Cambridge*)

180 acres of level ground about 5 miles from Cambridge was purchased. In July 1957 Sir Edward Appleton performed the opening ceremony at the new site and field laboratory, known as the Mullard Radio Astronomy Observatory. The principal radio telescope by 1958 was an aperture-synthesis interferometer comprising an E–W parabolic cylinder reflector 475 m long and 20 m wide, and a mobile reflector 58 m by 20 m which

could move along N–S railway tracks 300 m in length. The aerial system effectively synthesised an interferometer corresponding to two apertures, about 240 m by 150 m, separated E–W by 780 m. At $\lambda = 1 \cdot 7$ m, the system was equivalent to two aerials giving a primary beamwidth about 30′ divided into interferometer lobes separated by 8′. With the reflectors set at a given elevation, a declination band of width $4\frac{1}{2}°$ could be covered by 24 hours observing at each of 25 different positions of the movable aerial along its track. The output data was digitised and recorded on punched paper tape, and subsequently fed into an electronic computer to calculate the synthesised result. The system was successfully deployed in the succeeding years in the 4C survey of some 5000 sources.

Ryle's next scheme, initiated in 1962, was aimed at detailed surveys of particular regions of sky with very high resolution and sensitivity for the two-fold purpose of (a) mapping brightness distributions of radio galaxies and (b) making deep surveys of selected areas of sky to include very weak sources. For this purpose, Ryle employed equatorially-mounted steerable parabolic reflectors 60 ft in diameter, able to track continuously any given celestial region. The radio telescopes were placed along an E–W line, one being movable along a track, and synthesis was achieved in the following manner. With the rotation of the Earth the baseline covers in 12 hours all the directions in an elliptical ring as illustrated in Figure 4.3. By making daily recordings at different spacings, the synthesis fills a complete elliptical aperture. Since the maximum spacing between the radio telescopes was one mile, the system became known as the One-Mile Radio Telescope. To save observing time, Ryle's arrangement used three radio telescopes so that observations at two spacings could be made simultaneously, as well as at two frequencies, 408 MHz ($\lambda \approx 73$ cm) and 1407 MHz ($\lambda \approx 21$ cm). Synthesis of the elliptical aperture required 64 daily runs corresponding to different spacings, enabling a 3° region of sky to be mapped at a maximum resolution of 80 sec of arc at $\lambda \approx 73$ cm, and a 1° region of sky at a resolution of 23 sec of arc at $\lambda \approx 21$ cm. A new radio telescope, essentially on the same principle, but with more aerials, capable of operation at shorter wavelengths, and with a longer baseline, is under construction. It will have eight parabolic reflectors 42 ft in diameter operating

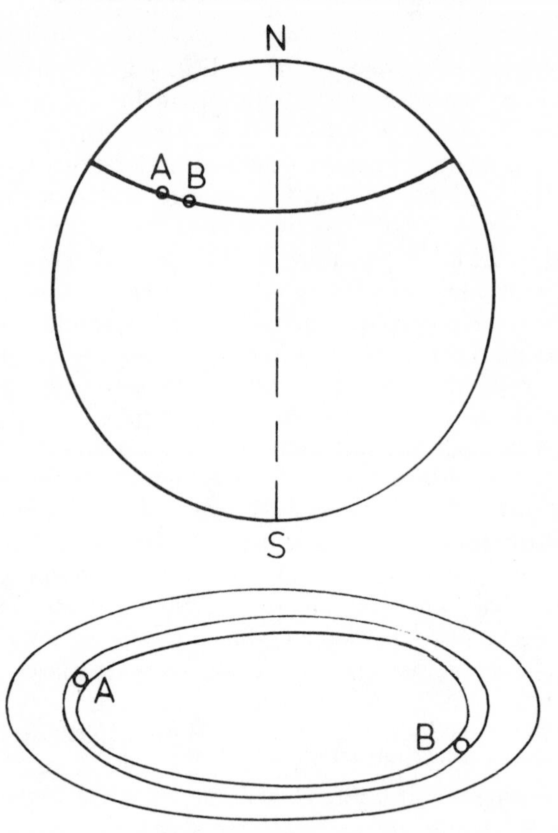

Figure 4.3 Aperture synthesis with the Cambridge One-Mile Radio Telescope.

down to $\lambda = 3$ cm spaced along a 5 km baseline, and the system should be completed in 1972.

The Cambridge programme has been characterised by the deployment of elegantly devised interferometer systems to achieve carefully planned objectives, and Ryle was knighted in 1966 for his contributions to radio astronomy. Ryle has been well supported by a very able team. In the design and analysis of aerial systems, A. Hewish, who became second Professor of Radio Astronomy at Cambridge in 1971, has always played an important role. Graham Smith left Cambridge in 1965 on his appointment as second Professor at Jodrell Bank.

Plate 4.4 The One-Mile radio telescope system at Cambridge (*by courtesy of M. Ryle, Cambridge*)

LONG BASELINE INTERFEROMETRY AT JODRELL BANK

Interferometers with horizontally separated aerials were soon being deployed at many radio observatories, for only in this way was it possible to construct instruments of really high resolving power, of the order of minutes or seconds of arc. It was primarily for the purpose of examining source structures of very small angular size that the Jodrell Bank group developed interferometer systems with very large spacings between the aerials, a research which has been led for many years with great foresight and skill by H. P. Palmer, although in the preliminary stages Hanbury Brown merits much credit for the originally proposed methods, while R. C. Jennison was prominent in the early experimental work.

When interferometer aerials are widely separated it becomes impracticable to connect them together with cables. At radio frequencies, attenuation becomes prohibitive over very long distances. One of the basic innovations to make long-baseline interferometry feasible was to introduce radio links to replace connecting cables. A convenient and versatile arrangement is to have a smaller, semi-mobile aerial at an outstation or temporary site, and to transmit the received signal to the home station to be combined with the signal at the main aerial, where delay lines are introduced as required to equalise path lengths. As a long radio link path near ground level is susceptible to atmospheric variations, phase stability cannot be maintained for long periods. Although the phase of the fringe pattern cannot be preserved, the amplitude of the fringes at various spacings provides a good indication of the main features of source sizes and structures. In fact, Michelson's stellar interferometer had relied solely on the assessment of fringe visibility, and in a similar way could differentiate between various simple models.

The endeavour in the early 1950s to determine the sizes and structures of radio sources such as Cygnus A stirred keen competition between the radio astronomy groups. In Australia, where the CSIRO group were by now exploring the potentialities of horizontally-spaced interferometers, Mills (1952, 3) designed a long-baseline interferometer with spacings up to 10 km, transmitting the radio frequency signals from one aerial to the other. In the first radio-link interferometer at Jodrell Bank, an entirely novel method was tried, by correlating the signals after they had passed through independent receivers and detectors at the two aerials. The transmission of post-detector signals over a radio link was a relatively simple matter. On combining the two signals, the resultant indicated the magnitude of the correlation function. The fact that the method worked at all by combining post-detector signals rather than radio frequencies was somewhat surprising. Hanbury Brown's inspired "hunch" successfully passed initial experimental tests on the Sun in 1950. The theory is rather complex and was not fully expounded until a few years later in a paper by Hanbury Brown and Twiss (1954). One snag was that the method could only work on input signals larger than receiver noise, a condition fortunately satisfied for the Cygnus and Cassiopeia sources which were the initial objectives when the full scale instrument using baselines up to 4 km was built by Hanbury Brown, Jennison and Das Gupta in 1952. Although the method fell out of favour in radio astronomy because the received power from most sources is too weak, the technique was successfully transferred to a new role, the measurement of the optical diameters of stars. The first successful optical test of the method was made by Hanbury Brown and Twiss at Jodrell Bank in 1956 on the star Sirius. Soon afterwards, Hanbury Brown left Jodrell Bank for the University of Sydney, Australia, to continue optical observations in a more favourable clime.

To cope with weak radio sources, the Jodrell Bank team now turned their attention towards the development of long-base interferometers in which radio frequencies are conveyed over radio links. In the attempts to bring down the lower limits of measurable size, interferometer baselines were continually being increased. As a consequence the fringes became very close together and the recorded fringe pattern moved so rapidly that it was necessary to introduce artificially a continuous phase shift to control the fringe speed. It was not, however, the first time that continuous phase-shifting had been applied, for Little and Payne-Scott (1951) at CSIRO had used the principle in constructing a swept-lobe interferometer to locate solar radio bursts. But the version introduced by Hanbury Brown, Palmer and Thompson (1955) incorporated new features both in frequency control and background elimination.

By 1962, Palmer's team at Jodrell Bank had lengthened the baseline to 60 000 λ at $\lambda = 1 \cdot 9$ m, enabling them to estimate angular sizes down to 1 second. The flexibility of the system was further improved by Rowson (1963) who introduced a continuously variable delay line to permit sources to be tracked across the sky with a changing projected baseline.

By this time I had resumed radio astronomy research with a small team at the Royal Radar Establishment, Malvern, where we had installed a variable-spacing interferometer with 25-metre radio telescopes. Further advances in long-base interferometry were now facilitated through a liaison with the Jodrell Bank team, when it was agreed to form an interferometer between a 25 m radio telescope at Malvern at one end, and the MK 1, or MK 2, radio telescope at Jodrell Bank at the other. This 127 km baseline, connected by a triple-hop radio link, was initially deployed at $\lambda \approx 21$ cm giving a separation up to 600 000 λ to measure angular diameters down to $0 \cdot 1$ sec of arc. Later by changing to shorter wavelengths, 11 cm and then to 6 cm, separations equal to 1 and 2 million wavelengths were achieved, giving resolutions of 0·05 sec of arc and 0·025 sec of arc respectively as described by Palmer, Gent, *et al* (1967). It was astonishing now to find radio instruments surpassing optical in resolving power.

Even with such long baselines, a proportion of the sources, or major components of them, remained unresolved; and in the late 1960s several research groups were contemplating a new way of operating widely-spaced interferometer aerials without a direct link between them. Such a possibility had been opened up by advancements in new techniques: firstly, the great stability of atomic clocks which could act as frequency controls for separate local oscillators; secondly, ultra high speed magnetic tape recording; thirdly, fast automatic computers to correlate the data from the two sites, when tapes were subsequently played back simultaneously. It was natural that Jodrell Bank should wish to take advantage of a scheme that would allow sites to be located anywhere, across continents or oceans, to achieve further extensions of baseline. Although Jodrell Bank developed their own version of the system, precedence in the successful exploitation of the new method was gained in 1967 by the Canadians, soon to be followed by the Americans. I shall return to a discussion of results of long-base interferometry later in Chapter 7.

To sum up the attitude to radio astronomical equipment at Jodrell Bank, it has essentially been two-fold—large single reflectors, and long-baseline interferometry. Through their basic simplicity and versatility, large steerable parabolic reflectors continue to be ideal types of aerial for many investigations. A steerable reflector of high precision, MK 2, an elliptical dish 120 ft by 80 ft was completed at Jodrell Bank in 1964. The 250 ft MK 1 has been undergoing modifications during 1971 to improve its surface and steering accuracy. Plans to construct a larger and more accurate reflector have suffered delays through lack of available funds, but there is every hope that the envisaged radio telescope, about 400 ft in diameter will be erected during the 1970 decade on a new site in Wales.

RADIO TELESCOPES IN AUSTRALIA

The Australian work on radio telescopes, which I shall now describe, is a story of great interest and importance to the development of radio astronomy. Much skill has been exercised in devising various arrangements of separated aerials near the ground in order to produce narrow beams. In fact, at one stage in the planning of future radio telescope systems at CSIRO, the choice between different alternative schemes led to a dramatic conflict of ideas within the group.

In 1952, B. Y. Mills decided to construct an ingenious aerial system to simulate a narrow beam by means of two long arrays lying N–S and E–W along the ground to form a cross. The two arrays producing two intersecting fan beams, were alternately connected together in and out-of-phase. The switched fan beams therefore alternately added and cancelled within the pencil beam formed at their intersection. Hence, by using a switch-frequency rectifier at the receiver output to respond only to signals at the switching frequency, the system effectively simulated a very narrow beam.[1] After a preliminary trial by Mills and Little (1953), the construction of a full scale "Mills Cross" with arms 1500 ft long was commenced in 1953 at Fleurs, near Sydney, and completed in the following year. Each array consisted of two rows of 250 dipoles, tuned to $\lambda = 3 \cdot 5$ m, backed by a plane wire-mesh reflector. The arrangement produced an effective beamwidth of 48′ and the direction could be altered by introducing different lengths of cable between the dipole elements. To simplify the procedure, only the N–S array was adjusted, allowing the beam to be pointed at different declinations, while the rotation of the Earth moved the beam around the sky. Used in this way, the Mills Cross proved to be a highly successful transit radio telescope. Its main disadvantage was that continuous steering was impracticable; also, sidelobes tended to be rather strong, although they could be reduced by tapering the response of the arrays. A few years later at CSIRO a similar Cross was constructed by Shain at 15 m wavelength. The subsequent building in several countries of various types of Mills Cross is itself a tribute to the value of this novel method of obtaining a narrow beam.

A large part of the Australian programme has always been devoted to the Sun, and various aerial systems have been designed specifically for solar observations. The Sun presents a

[1] Ryle (1952) had mentioned this advantage arising from the use of long arrays, N-S and E-W respectively, in a phase-switching interferometer.

unique type of problem, because it stands out as a conspicuous, large source with the dual requirements of (a) mapping the distribution of radio brightness of the quiet Sun and the slowly-varying features like sunspots (b) observing and tracking transient phenomena associated with solar flares. To determine the intensity distribution across the quiescent Sun, Christiansen (1953) designed a multi-element, spaced array which produced a set of beams. The first system comprised 32 parabolic dishes, of 6 ft diameter, spaced at regular intervals along a straight line 710 ft long. This multi-element array, sometimes called the grating interferometer, produced at 21 cm wavelength a series of fan beams 3′ wide and separated by 1·7 degrees, as illustrated in Figure 4.4. The Sun, subtending half a degree could then be observed as it moved through each beam in turn. A fan beam, of course, observes a strip across the face of the Sun. Later, Christiansen and Mathewson combined the grating array with the Mills Cross principle. In this version, sometimes called the 'Chris-Cross', two lines of aerials running N–S and E–W, each consisting of 32 parabolic reflectors 19 ft in diameter, were

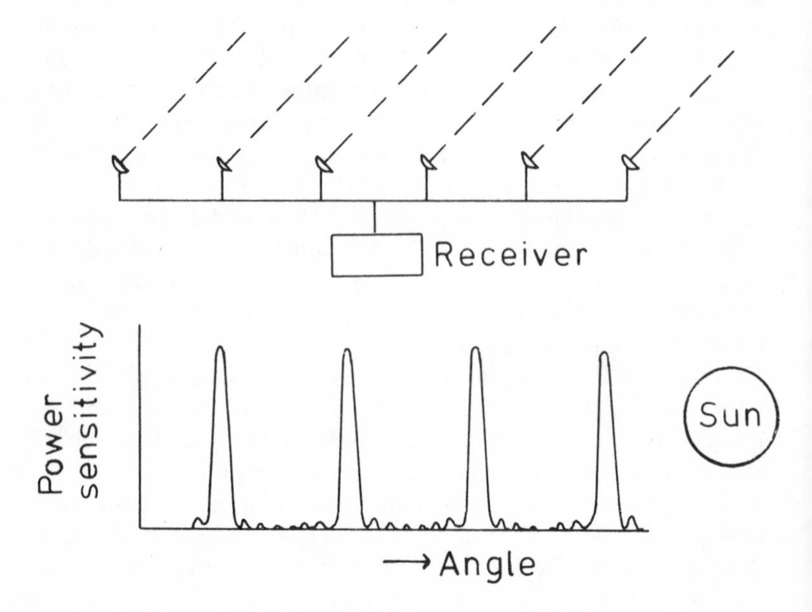

Figure 4.4 Principle of the grating interferometer.

switched in and out-of-phase. The intersecting sets of fan beams produced a matrix of pencil beams, 3′ wide, spaced 1° apart. As the Sun drifted through the succession of beams a detailed picture of the radiation intensity over the face of the Sun could be derived.

The investigation of bursts of radio emission associated with solar activity presented a quite different task. Such bursts can vary rapidly, in seconds or minutes, in strength, position and spectrum. To catch these transient phenomena, specialised methods had to be devised. To record rapid changes in the spectrum, Wild and McCready (1950) developed a radio spectrometer consisting of an aerial able to accept a wide band of frequencies connected to a receiver with its tuning rapidly sweeping over the band. And to determine the position and movement of the transient source, Little and Payne-Scott (1951) devised an interferometer with the lobes made to move rapidly across the Sun by introducing a rotary phase changer. When it became clear that many types of burst altered rapidly in spectrum and position, Wild and Sheridan in 1957 developed a swept-frequency interferometer. The pursuit of solar radio astronomy by CSIRO was highly successful and crowned by the emergence of an orderly analysis and portrayal of solar radio phenomena.

Bowen and Pawsey had collected together at CSIRO an extremely able group. In addition to the many important CSIRO contributions which I mention in other chapters, it is important not to overlook the theoretical studies of aerial systems and the resolution that can be achieved. Bracewell (1954, 8) in particular made several excellent basic analyses of aerial problems.

In the early 1950s, radio astronomy had become thoroughly established as an outstanding fruitful enterprise with a multitude of research problems clamouring for attention. The situation at CSIRO is vividly illustrated in the following informal remarks by Wild:

"I can well remember the very special atmosphere that existed in the Lab at that time—perhaps, though on a humbler scale, it was something like the atmosphere at the Cavendish after Rutherford's discovery. New results were popping up all the time. Almost everything you did led to a discovery. About once per fortnight the research staff would meet together with Pawsey in the chair. Each person and team would report their progress, and Pawsey would express his approval or surprise or criticism of each in turn. You might possibly imagine that complete harmony existed. Nothing could be farther from the truth—there was a great deal of squabbling whenever interests of different teams overlapped. This was partly the result of healthy competition, and partly perhaps due to individuals assuming the role of prima donna."

It was clear that to tackle the problems that lay ahead, and to keep in the front rank of world research, demanded new radio telescopes, but how to finance costly new projects was the real obstacle. The very qualities which had contributed so much to the success of the Australian group, the persistence, independence and foresight of individual members, now inevitably drew them into a cleavage of competing priorities. Funds were insufficient to support all the proposals. The principal contenders were Bowen who wished to have a large steerable dish capable of operating down to short wavelengths, Wild who was planning a complex aerial system to synthesise radio pictures of the Sun, and Mills who proposed a larger and more elaborate version of the Mills Cross.

Bowen, as head of the Radiophysics Lab., and Pawsey as radio astronomy leader, had the responsibility of deciding which scheme should have precedence. Bowen wanted a steerable dish comparable in size but more accurate than the MK 1 radio telescope at Jodrell Bank. He was convinced of the desirability of having a versatile, general purpose instrument, able to operate over a wide range of wavelengths, and readily adaptable to different tasks as new discoveries were made. Pawsey, on the other hand, was more attracted by the special merits of interferometry and the prospect of ingenious combinations of aerials to provide fine angular resolution. What one could be certain of was Pawsey's unswerving loyalty to CSIRO, so that when the final decision had been taken there would be no doubt of his unqualified acceptance. Nevertheless, a storm of contention had broken over the group with repercussions that caused a split, ultimately to be resolved only after movements of staff and a measure of reorganisation had taken place.

Plate 4.5 E. V. Appleton in discussion with J. G. Bolton (right) (*by courtesy of the Radiophysics Laboratory, CSIRO, Australia*)

Bolton was one of the staunch supporters of Bowen's proposal for a large parabolic reflector. After Bolton had exhausted the scope of the cliff edge interferometer with the notable measurements he had accomplished on discrete sources, he transferred for a time to Bowen's team working on cloud physics, until in 1955, with Bowen's connivance, Bolton was given the opportunity to be the first director of the Owens Valley Radio Observatory of the California Institute of Technology. Here the President of the Institute, L. DuBridge, was keen to establish a radio astronomical observatory that would match the nearby great optical observatories of Mt. Wilson and Palomar. In 1955, Bolton assisted by his colleague Stanley, moved to Cal. Tech. to take charge of the radio observatory.

By 1956 Bowen had succeeded in raising the necessary funds to fulfil his intention to construct a large steerable telescope for CSIRO. Realisation of the project was made possible through generous donations; the Carnegie Corporation of New York and the Rockefeller Foundation contributed altogether $607 000; private subscribers within Australia gave £27 000; and the balance, rather more than half the total cost of £650 000 for

Plate 4.6 E. G. Bowen (left) and J. L. Pawsey (right) discuss the plans for the 210 ft CSIRO radio telescope with L. Biermann of the Max Planck Institut (*by courtesy of the Radiophysics Laboratory, CSIRO, Australia*)

the telescope and site facilities, came from the Australian Government. A detailed specification was not prepared at the outset, but evolved gradually after the engineering implications of several possible designs had been worked out by the consultant engineers, Freeman, Fox and Partners of London. During the study phase, Minnett of CSIRO was attached to the consultants to ensure continuous collaboration. By 1957, Freeman, Fox and Partners had prepared a design that would meet the CSIRO requirements within the stipulated cost. All was now decided; the reflector would be 210 ft in diameter, with surface deviations less than 9 mm at any orientation, and pointing accuracy within 1 min. of arc. The telescope was to be fully steerable on an altazimuth mount, the only concession being a minimum elevation of 30°, thus lowering the height required for the supporting tower. Barnes Wallis, the famous aircraft designer, had been consulted on the problem of accurate

Plate 4.7 The CSIRO 210 ft radio telescope at Parkes, Australia (*by courtesy of the Radiophysics Laboratory, CSIRO, Australia*)

steering, and he devised a novel master unit to control the direction of the telescope. Tenders for construction of the telescope were invited from American, German, and British firms, and by 1959 the main contract was awarded to the West German firm, M.A.N. Two years later the telescope was completed at Parkes, 200 miles west of Sydney, and admirably fulfilled expectations with its performance maintaining optimum efficiency down to 10 cm wavelength. Bolton by this time had returned from Cal. Tech. to take charge of the operation of the

Parkes radio telescope and its research programme, while Stanley remained to assume the directorship at Owens Valley.

Meanwhile, Wild and Mills put forward their proposals for other types of radio telescopes. As it was felt that the lead in solar radio astronomy held by CSIRO could not be abandoned,

Plate 4.8 F. G. Smith examining solar radio spectrograph transmission lines with J. P. Wild (right), H. C. Minnett (left) and S. Chapman (right centre) (*by courtesy of the Radiophysics Laboratory, CSIRO, Australia*)

priority was accorded to Wild's scheme for a radio spectroheliograph. Even for this instrument, no funds were then available, so the prospects for a new "Super Cross" proposed by Mills seemed extremely bleak. It is evident in letters written by Bowen at this time that he considered the "Super Cross" an excellent scheme, but reluctantly he felt forced to admit that a third large new radio telescope was, for some years at least, quite beyond the financial resources that CSIRO could possibly hope to muster.

Unforeseen circumstances now intervened to mould the course of events. At the University of Sydney, the Professor of Physics, H. Messel, had appointed C. S. Gum to develop an optical observatory. But Dr. Gum's untimely death in a skiing accident while on holiday in Switzerland brought a sudden halt to the project. An alternative idea now occurred to Messel. He would offer the accumulated funds (about $200 000) to Mills to construct his new Cross and start a radio astronomy group at the University of Sydney. Mills readily seized the opportunity. At the same time, Christiansen accepted the invitation to become Professor of Electrical Engineering at the University, so the pooling of resources of the departments was clearly an advantageous prospect. In 1960 Mills moved to Sydney University and began to formulate the detailed design for the new Cross. He decided to construct the arms in the form of mile-long cylindrical parabolas with mesh reflectors 40 ft wide, with lines of dipoles at the focus for a principal frequency of 408 MHz ($\lambda \approx 73$ cm) giving a beamwidth of 2·8 min of arc, and a subsidiary frequency of 111·5 MHz ($\lambda \approx 2{\cdot}7$ m) giving a beamwidth of 10 min of arc. By dividing the N–S arm into sections connected to preamplifiers, and joining them together in different phase relationships, observations at 408 MHz could be made at 11 separate beam positions simultaneously. The instrument was to be operated as a transit telescope, altering the beams in elevation as required and utilising the Earth's rotation to survey the sky.

Such a large and complex instrument obviously needed more funds than the amount initially available. Fortunately the National Science Foundation of America responded to the appeal for financial aid. When the Cross was completed at a site at Molonglo, near Canberra, in 1967, about $900 000 had

Plate 4.9 B. Y. Mills with the E–W arm of the Mills Cross, Molonglo in 1964 (*by courtesy of B. Y. Mills, Sydney University*)

been contributed by the National Science Foundation. By 1970, a further improvement had been made to the Cross by arranging two extra fan beams from the E–W arm, so producing a total raster of 33 separate beams. The Molonglo Cross, as it is now called is shown in Plate 4.10.

While the negotiations for the Cross were proceeding at Sydney University, plans were launched for the radio heliograph at CSIRO. By 1962, the Ford Foundation, USA, responded to appeals from Bowen and Pawsey by making a generous

Plate 4.10 The Mills Cross at Molonglo, Australia (*by courtesy of B. Y. Mills, Sydney University*)

grant of \$630 000, the estimated amount required to embark on the project to produce pictures of solar radio activity. Wild planned to achieve this objective with 48 beams, each 3·5 min arc wide at 80 MHz ($\lambda = 3{\cdot}75$ m) scanning rapidly across the Sun in order to form pictures of the radio Sun on a television screen. A pair of pictures, in two polarisations, was to be produced each second. The aerial system was to consist of 96 steerable parabolic reflectors, each of 13 m aperture, disposed around a circle of 3 km diameter. To form multiple, independent beams, the 96 aerial units and preamplifiers were all connected to a central control room, where, with the aid of a complex network of delay lines, interconnections, and a computer, the multiple beams and rapid scanning could be generated. A new site was chosen at Culgoora, 300 miles NW of Sydney, where the radio heliograph was installed, together with a new radio spectrograph sweeping from 5 MHz to 8000 MHz. The Culgoora installation came into operation in 1967, the same year as the new Mills Cross at Molonglo.

The two arrays, at Culgoora and Molonglo, possess certain common principles. Both comprise an assembly of aerial units having the relative spacings that are to be found in an aperture of similar maximum dimensions. This fulfils the key requirement for the synthesis of single narrow beams. By incorporating a complex and flexible interconnecting system, a number of separate beams can be formed in different directions at the same time. This constitutes a step towards an ideal, simultaneous aperture synthesis to produce a matrix of effectively independent narrow beams.

Both these great installations closely conformed to Pawsey's own ideas of the most desirable developments in radio telescopes. He gave Mills moral encouragement, as well as being more directly concerned in supporting Wild's scheme at CSIRO. By 1962, Pawsey felt that his ambition, to establish front-rank radio astronomy research in Australia, had been achieved. By now, former junior team leaders were being promoted to senior positions. The time seemed opportune for Pawsey to step aside to allow leading members a fuller independence. Pawsey was therefore about to accept an invitation to become director of the recently established National Radio Astronomy Observatory, USA, when he became ill, and colleagues and friends throughout the world were shocked to learn there was little hope of his recovery. Pawsey resumed his duties in Australia and with characteristic devotion, during prolonged illness, he completed various tasks, one of the last being to edit a Radio Astronomy Issue of the Australian Proc. I.R.E. which included papers on the new developments in Australia, the 210 ft radio telescope, the radio heliograph, the Sydney Cross, as well as descriptions of advances in many parts of the world, representing an apt memorial to a dedicated pioneer of radio astronomy. Pawsey's death seemed all the more poignant following so soon on the struggle of interests and priorities at CSIRO. But it would be a mistake to think such conflicts did not confront other groups. They are an inevitable accompaniment of an expanding, exciting field of research involving costly equipment. The CSIRO story dramatically highlights problems experienced in some degree elsewhere. The splitting away of some members as they advance in seniority and seek new ventures happens to many groups. And the question of how to finance new large radio telescopes often becomes a national concern, as witnessed for instance by Radio Astronomy Committees set up in USA and Great Britain. All the Australian radio telescopes have been highly successful, and some of the important achievements are described elsewhere in this book. At the CSIRO Division of Radiophysics, Bowen's strong leadership of the Division, his support and concern for radio astronomy, has been a major influence; on his retirement in 1971, Wild was appointed his successor as Chief of the Division.

RADIO ASTRONOMY AT THE NAVAL RESEARCH LABORATORY, USA

In the five years or so following the war, scientists in USA on the whole were slow to recognise the potentialities of radio astronomy. It was surprising that with their scientific and engineering resources they did not seize more readily the opportunities presented by the initial discoveries. In 1948 Reber endeavoured to elicit support for a design he proposed for a 220 ft radio telescope, but he was too far ahead of any official interest in such a possibility. In his words ". . . for the most part the attitude was that I was harmless and if no interest was shown in what I was saying, I would go away quietly". In these years, apart from the completion in 1946 of the US Army Signal Corps project to detect radar echoes from the Moon, the only important new work in radio astronomy in USA was initiated by the US Naval Research Laboratory in Washington. Here the research started by J. P. Hagen, along with F. T. Haddock and others at NRL, represented a highly significant step, leading the way in short wavelength radio astronomy, and strongly influencing the pattern of much subsequent research in USA. Their 50 ft parabolic reflector, completed in 1950, was the first large radio telescope built specifically to operate at wavelengths down to 1 cm.

Radio astronomy has owed much to industrial, military, and government research establishments, where the quality and outlook of research staff has often led them to explore basic fields in which equipment and techniques devised in the laboratories can be exploited for research. The establishments generally have taken a beneficent attitude to such activities, recognising their value in promoting scientific status and contributing to scientific advance. Although astronomical observations can usefully provide a testbed for new devices, the pursuance and completion of astronomical research projects demands sustained and dedicated effort, far beyond the needs of experimental tests. Nevertheless, new observing techniques may reveal unexpected phenomena, and progressive establishments like to encourage research associated with the equipment they develop.

The trend during the war in radar and radiometry, as well as in radio communication had been towards the increasing exploitation of short wavelengths. In 1945 the application of short centimetric and millimetric waves was in its infancy, and the promise of high resolution radars producing well-defined pictorial displays of ships, coastlines, and so on, offered an attractive prospect. Navigation by radio emission from the Sun was amongst other possible applications, and the Collins Radio Company developed under a US Navy contract a radio sextant utilising tracking of the Sun at about 1 cm wavelength. With such a background of interests it is natural that NRL should be concerned with research and applications at cm and mm wavelengths, and this foundation provided a fertile ground in which the radio-astronomical initiative of Hagen and Haddock could flourish. Their first investigations were mainly directed to the interpretation of measurements of solar radiation, including eclipse observations, at wavelengths between 8·5 mm and 9·4 cm, using parabolic reflectors up to 10 ft diameter. Meanwhile the construction of the 50 ft reflector proceeded, the final design and manufacture being carried out by the Collins Radio Company with N. L. Ashton of the University of Iowa as consultant. The reflector was composed of 30 aluminium sector castings which were bolted together, and the surface was then machined to parabolic shape within a tolerance of ± 1/32 inch, in a boring mill specifically set up for the job. The reflector segments were then dismantled and reassembled on an altazimuth mount consisting of a modified naval gun bearing placed on a laboratory roof at NRL. After re-erection, the final errors did not exceed 1/16 inch and even these deviations were confined to small regions of the reflector. This excellent radio telescope, shown in Plate 4.11, has been instrumental in making some of the most fundamentally important observations in

Plate 4.11 The 50 ft reflector at NRL, Washington (*by courtesy of C. H. Mayer, Naval Research Laboratory, USA*)

centimetric radio astronomy, including first detections of emission nebulae, radio emission from the planets Venus, Mars and Jupiter, polarisation in discrete sources, hydrogen line absorption, as well as centimetric radar echoes from the Moon; and it was the first radio telescope to be fitted with a low-noise maser receiver.

From 1954 Hagen worked for several years on the space rocket development programme, later resuming microwave studies of the Sun in 1966 when he returned to radio astronomy research at the University of Pennsylvania. In 1956 Haddock was appointed head of a radio astronomy observatory at the University of Michigan, acquiring there an 85 ft precision radio telescope with financial support from the Office of Naval Research. McClain became leader of the NRL group, and an 85 ft precision radio telescope, with equatorial mount, was erected at NRL's Maryland Point Observatory, superseding the 50 ft reflector both in collecting area and pointing accuracy. In

Plate 4.12 The NRL team on a solar eclipse expedition in 1950. Left to right back row: H. Herman, D. S. Hawkins, T. P. McCullough, J. P. Hagen, F. Haddock. Front row: R. M. Sloanaker, E. Beck, C. H. Mayer, D. R. J. White (*by courtesy of C. H. Mayer, Naval Research Laboratory, USA*)

1968 Mayer, who had been a member of the NRL radio astronomy team since its inception succeeded to the leadership. Practically all the research by Hagen, Haddock, McClain, Mayer and their colleagues has throughout the years continued to be particularly notable for impressive achievements at cm and mm wavelengths.

One disappointing experience in which NRL became partly involved was the US Navy project to build a steerable 600 ft radio telescope, which was abandoned after construction had commenced on a site at Sugar Grove, West Virginia. Although such a large reflector could not be expected to function at centimetric wavelengths, it was envisaged that sufficient accuracy could be attained to operate down to the wavelength of the 21 cm hydrogen line. A huge steerable parabolic reflector would have many applications, such as investigations of ionospheric phenomena, radio communication, tracking of space probes, as well as radio astronomy, and it was planned to make the instrument available to universities and institutions in addition to NRL. Two groups at NRL had a direct interest, the radio astronomy group led by McClain, and Trexler's communication engineering group who were examining radio communication via lunar reflections. By 1959 preparation of the site and the preliminary stages of construction of the world's largest steerable radio telescope had begun. Unfortunately, the project was overambitious, and the lessons that might have been learned from the struggles at Jodrell Bank had been ignored, for it was not sufficiently appreciated that the same predicament of escalating costs and engineering problems would be equally likely to beset this vaster enterprise. To quote from a 1962 statement, "In the past year with only the foundation laid and control stations built, construction has come to a virtual halt while the 20 000 ton paraboloidal antenna was being redesigned. Considerable engineering difficulty was encountered in designing the giant dish, for the shape of its surface was to remain perfect within a fraction of an inch, despite stresses of wind, motion and temperature, while the telescope was pointed anywhere in the sky. In 1958 the total cost was estimated at 79 million dollars. When this figure rose to more than 200 million, the Secretary of Defense ordered the project cancelled although some 96 million dollars had already been spent".

THE USA NATIONAL RADIO ASTRONOMY OBSERVATORY

In 1954, the awareness that in radio astronomy the United States was seriously lagging behind other countries, particularly Britain, Australia, Holland and Russia, was expressed at a conference held in Washington in January of that year. It was also realised that a major financial issue was involved, since the large instruments that would be needed at this stage for an advance to the forefront of radio astronomy could be well beyond the resources of any single institution. D. H. Menzel, Director of Harvard Observatory, had already been preparing, along with B. J. Bok of Harvard and J. B. Wiesner of MIT, joint proposals for the expansion of radio astronomy facilities. Realising that in the long run it would be more advantageous in the interests of all to establish a national radio astronomy observatory with the finest possible equipment, available to any university or institutional radio astronomy research group, Menzel supported by other optical and radio astronomers decided to approach L. V. Berkner, president of Associated Universities Incorporated (AUI), whose function it was to recommend and organise large scale research facilities on behalf of the universities. Accordingly, Menzel presented a memorandum entitled "Survey of the potentialities of cooperative research in radio astronomy", the outcome being that AUI agreed to the preparation of detailed plans with a view to submitting an application to the National Science Foundation for financial support. The complicated task of preparing plans for the National Radio Astronomy Observatory commenced with the appointment of R. M. Emberson as acting director of the project and a steering committee under J. P. Hagen. Many competing proposals were examined, and the projected scheme had to include all aspects, radio telescopes, siting, staffing, laboratory facilities and organisation. Eventually it was agreed that the principal instrument should be a large radio telescope capable of centimetric operation. After a lengthy consideration of possible parameters it was decided to specify a parabolic reflector of 140 ft diameter on an equatorial mount, and to entrust the design to N. L. Ashton, who had been mainly responsible for the design of the 50 ft NRL reflector.

It was perhaps hardly surprising, in countries that already held supremacy in radio astronomy research, to hear, in conversation at least, certain critical comments on the American proposal for a national observatory. Some thought it a grandiose scheme attempting to promote research by organisation and committees, instead of fostering facilities around a nucleus of individual scientists who had proved their worth with meagre equipment. Now, in 1972, few would query the merits of both methods of approach and that the choice depends on circumstances. In Britain, for instance, with scientists of the individuality and calibre of Lovell and Ryle it would have been unthinkable not to support both independently. But no one now doubts the wisdom of the American plan with NRAO successfully established with first class equipment shared by many universities in addition to having its own research staff. The outpouring of important scientific papers emanating from NRAO is ample testimony to the soundness of the initial concept.

Like many radio observatories, however, NRAO had to pass through a number of tribulations before the original plan was fulfilled. By mid 1959 the astronomer Otto Struve had been appointed director of NRAO, an 85 ft radio telescope had been erected at the site at Greenbank, West Virginia, and a research programme was under way. But difficulties were already being encountered in the construction of the 140 ft radio telescope, for the polar shaft and yoke became a centre of anxiety when it was

Plate 4.13 The 140 ft NRAO radio telescope at Green Bank, USA (*by courtesy of the National Radio Astronomy Observatory, USA*)

found they were subject to brittle fracture under stress. Some engineers had previously expressed doubts about the mechanical feasibility of a polar mount for such a large reflector. Delay ensued while a solution was sought to overcome this engineering hurdle of material strength, and finally, in 1965, the 140 ft telescope was successfully brought into operation.

The delay was not without compensation. Largely due to the efforts of J. W. Findlay, a large parabolic wire-mesh reflector was built to meet the demands of the interim period. Steering in elevation only, this 300 ft transit telescope suitable for wavelengths of 21 cm and above, performed yeoman service after its completion in 1962. Other instruments have since been added including an interferometer comprising three 85 ft radio telescopes with variable spacing up to a separation of 2·7 km. Another notable recent addition to the NRAO equipment, indicative of the awakening interest in still shorter wavelengths, is the 36 ft diameter mm wave reflector at Kitt Peak, a high site

Plate 4.14 The 36 ft NRAO mm wave radio telescope at Kitt Peak, USA (*by courtesy of the National Radio Astronomy Observatory, USA*)

deliberately chosen to minimise atmospheric attenuation and fluctuation.

It was unfortunate that Struve's period of directorship should have been burdened by the worries of constructional delays and difficulties, and for health reasons he relinquished the office in 1962. He would have been succeeded by Pawsey but for the tragic illness which prevented his taking up the post. Subsequently D. S. Heeschen was appointed director and during his leadership the observatory has come into the full operation originally envisaged.

There is no doubt the National Radio Astronomy Observatory with its progressive acquisition and utilisation of excellent radio telescopes now ranks as one of the finest radio astronomy centres in the world. Plans are going ahead for the construction of a very large array (VLA) comprising 27 parabolic reflectors, each 25 m (82 ft) in diameter, distributed along three arms of 21 km length in the form of a Y, with all the aerials mobile along railway tracks. The system will enable observations to be made at multiple spacings simultaneously, and by 12 hours of observation as the Earth rotates it will be possible to synthesise radio source maps comprising 10 000 picture elements of 1″ resolution at $\lambda \approx 11$ cm, and still higher resolution at shorter wavelengths. The proposed site in New Mexico combines the advantages of a near equatorial site for optimum sky coverage together with a low interference location. The present estimated cost of the system is 76 million dollars.

The observatories in Britain, Australia, and the USA so far described illustrate the problems encountered in establishing major radio-astronomical research centres. I shall now discuss more briefly some other great radio observatories, but the space allocated in no way reflects their importance. The simple fact is that a whole book could be devoted to the history of observatories, and in this short account I cannot do justice to them.

RADIO TELESCOPES IN THE NETHERLANDS

The pursuit of radio astronomy by the Leiden astronomers culminated in the erection at Dwingeloo of a 25 m (82 ft) radio telescope of a unique and economical design. As early as 1946

Oort had consulted R. J. Schor, a leading designer and director of the firm of Werkspoor, and a preliminary scheme was formulated. With the approval of Minnaert and other astronomers, a Netherlands Foundation for Radio Astronomy was inaugurated to sponsor recommendations.

The Prime Minister, Prof. Dr. Ir. W. Schermerhorn, who was himself a meteorological scientist, had a sympathetic interest in fostering research. In this reassuring atmosphere, and in a country which has always held a pre-eminent place in astronomy, Oort's scheme was granted financial backing by the Netherlands Organisation for the Advancement of Pure Research.

A design was evolved which incorporated several original ideas. The wire-mesh reflector consisted of 1464 plane facets in sets of four which could be separately adjusted to the correct profile position after mounting on the reflector framework. In this manner, the whole surface was set close enough to parabolic shape to be well within the tolerance for 21 cm wavelength. In fact, when Werkspoor later built for my group at RRE, Malvern, an identical reflector with finer mesh over the central region, it operated near optimum efficiency at 10 cm wavelength. An engineering consultant, B. Hooghoudt supervised the Dwingeloo project which was completed in 1956. At the time of erection, it was the world's largest fully-steerable radio telescope.

The next scheme conceived by Leiden was an elaborate form of Mills Cross to give a 1 min of arc beam at $\lambda = 21$ cm. As the proposal seemed likely to be expensive, it was agreed to make the venture a joint project of the Benelux countries, Belgium, Luxembourg, and the Netherlands. The Cross system, devised by Christiansen (from Australia), Erickson, and Högbom, consisted of 100 equatorially-mounted reflectors with preamplifiers, which were to be connected together in a variety of phase relations to provide up to 100 beam positions simultaneously. The Benelux Cross did not materialise, it was too great a leap forward, it was too costly and complex. So the Netherlands substituted a simplified plan of its own comprising twelve 25 m reflectors along an E–W line 1·6 km in length, with the two end dishes movable on rails. This new installation, at Westerbork, came into operation in 1970.

Plate 4.15 The 25 m radio telescope at Dwingeloo in the Netherlands (*by courtesy of H. C. van de Hulst, Leiden*)

RUSSIAN RADIO TELESCOPES

In USSR, radio astronomy has long been a flourishing branch of research. In this sphere of activity a friendly atmosphere has been maintained with all countries, and there have been no barriers except that of language. Fortunately, the problem of the incomprehensibility of the Russian language to most Western scientists is rapidly being overcome by the increasing availability of translations.

Most of the Russian radio astronomy has centred on the Lebedev Institute of Physics, Academy of Sciences, Moscow. Many of the USSR radio telescopes have been designed by P. D. Kalachov of the Radio Astronomy Laboratory directed by V. V. Vitkevich. The largest radio telescope is a Mills Cross system at Serpukhov, near Moscow, with arms in the form of parabolic cylinder reflectors, 1000 m long and 40 m wide. At very short wavelengths S. E. Khaikin was responsible for initiating several important radio telescopes, the best known being the 22 m parabolic reflector designed by Kalachov and Salomonovich for operation down to 8 mm wavelength. This excellent instrument, when completed at Serpukhov in 1959, was the world's finest mm wave radio telescope. A second radio telescope of the same type has since been erected in the Crimea.

Plate 4.16 The 22 m radio telescope at the Lebedev Physical Institute, near Moscow, (*by courtesy of P. D. Kalachov, Lebedev Institute, Moscow*)

A notable design has been the multi-plate system developed by Khaikin and Kaidanovsky, and built at Pulkova, near Leningrad, consisting of 90 tiltable plates 3 m by 1·5 m placed along an arc and adjusted to reflect the radio waves to a common focus. The relatively small reflector plates supported near the ground ensured accuracy and stability, and at $\lambda = 3$ cm the system gave a fan beam of width 1′ by about 10′ (a value depending on the elevation setting). Such an arrangement is of course essentially a transit telescope, utilising the rotation of the Earth to scan the sky. Construction of a larger version is in progress with reflectors covering a complete circle to enable the beam to be directed to any azimuth.

KRAUS RADIO TELESCOPES IN USA AND FRANCE

The building of long reflectors of limited steering, with the long dimension parallel to the ground, is an attractive idea for the economical construction of large, partially steerable radio

Plate 4.17 The Kraus two-reflector radio telescope at Ohio, USA (*by courtesy of J. D. Kraus, Ohio State University*)

telescopes. One of the best known, constructed at Ohio State University, USA, in 1962 was designed by J. D. Kraus, who was already a recognised authority on radio aerials. The radio system, shown in Plate 4.17 consists of a tiltable flat reflector made of wires stretched on a metal frame, 340 ft by 100 ft, which reflects radio waves from the source on to a fixed section of parabolic reflector 360 ft long by 70 ft high, and thence to a horn at the focus. The ground is covered with aluminium sheet which serves the dual purpose of simplifying the horn design and preventing the reception of thermal radiation from the ground. The system provides a transit radio telescope with a fan beam, about 5′ by 40′ at $\lambda = 21$ cm, adjustable in elevation by tilting the flat reflector.

The ingenuity and success of the Kraus aerial led the French radio astronomers to copy the method in a larger, improved version built at Nancay, France. Here the curved reflector is 300 m by 35 m, and the tiltable one 200 m by 40 m; and the curved reflector has a spherical profile instead of parabolic so allowing some alteration of beam direction in azimuth by moving the receptor at the focus. Radio astronomy has been actively pursued in France since 1947, and the largest group led by Denisse is based on Meudon Observatory and has its main observing station at Nancay. For many years, research concentrated on solar phenomena; for this purpose, interferometers were set up at wavelengths of 3 cm and 21 cm as well as a multichannel Cross of parabolic reflectors. With the completion of the large Kraus-type radio telescope the emphasis of the work has shifted towards galactic and extragalactic radio astronomy.

RADIO TELESCOPE INTERFEROMETERS AT CAL. TECH. AND RRE

In the mid 1950s, plans put forward independently by the California Institute of Technology, USA, where Bolton became first director, and by my group at the Royal Radar Establishment, England, marked a significant step in the development of radio telescope systems. We both realised the advantages in constructing interferometers with two equal, large, steerable parabolic reflectors movable along rails to vary the spacing.

With such a system, sources could be tracked continuously, positions determined with great accuracy, and structures derived. We both planned to use parabolic reflectors of about 25 m diameter which we knew could be constructed without undue difficulty for wavelengths down to 10 cm. The Cal. Tech. installation was in operation at Owens Valley in 1957 with two equatorially-mounted 90 ft reflectors which could be placed at a number of spacings up to 1600 ft along an E–W baseline, and was soon proving its worth by a striking series of source measurements. During the 1960s, the N–S baseline was added, also a 130 ft radio telescope with altazimuth mount.

The installation at RRE, Malvern, completed in 1961, comprised two 25 m reflectors with mobile mounts on intersecting railway tracks permitting any spacing up to about 1 km in any orientation; in addition to the receivers, various high power transmitters were fitted for upper atmospheric research.

Many subsequent interferometer and synthesis systems have incorporated in various ways steerable, mobile, parabolic reflectors of a convenient, moderately large diameter. Such arrangements are versatile and adaptable, and lend themselves to later expansion with additional units as funds become available.

RADIO TELESCOPES AT NRC, CANADA

Perhaps the longest sustained centimetric solar watch has been that maintained by Covington of the National Research Council (NRC) of Canada. Starting in 1946 with a simple parabolic tracking reflector, Covington later designed an interesting type of compound interferometer producing a single narrow fan beam, about 2° by 1′, and in a subsequent version a width of only 15″. In the years after the war, McKinley at NRC was engaged on another aspect of radio astronomy, radar echoes from meteor trails. In the 1960s, a reorganisation and expansion of the radio astronomy facilities at NRC culminated in the Algonquin Radio Observatory with McKinley as director, and the construction of a 150 ft altazimuth radio telescope designed by Freeman, Fox and Partners, provided NRC with one of the most accurate large radio telescopes in the world, similar in performance to the NRAO 142 ft reflector, and able to operate at wavelengths down to 2 cm.

LARGE RADIO TELESCOPES IN USA

There are more large radio telescopes in USA than in any other country. Not only is America richly endowed with radio astronomy observatories, but there are also large radio telescopes primarily intended for ionospheric research, space research, or military research that are available for research projects in radio astronomy. One example is the great reflector at the Arecibo Ionospheric Observatory completed in 1963, operated by Cornell University, and funded by the Advance Research Projects Agency of the Department of Defense. The scheme was initiated, following a suggestion by W. E. Gordon of Cornell University, for studies of the scattering of radio waves by electrons in the ionosphere. It is a most extraordinary radio telescope, a wire-mesh spherical reflector 1000 ft in diameter supported over a great natural hollow in the ground. Steering up to 20° from the zenith is achieved by moving the feed at the focus, and the equatorial location of the site is especially appropriate for planetary observations and measurements of discrete sources by lunar occultations.

At the Lincoln Laboratory, MIT, the 120 ft reflector known as Haystack, completed in 1964 with financial support from the US Air Force, was designed as a research radar operating at wavelengths down to 1 cm. It is able to retain its precision in profile with the aid of a surrounding radome protecting it from wind and weather. This radio telescope has been instrumental in making some of the finest planetary radar observations obtained from a terrestrial site. The Jet Propulsion Laboratory's 210 ft Goldstone reflector, built for tracking space vehicles, is another example of a large accurate radio telescope which has figured prominently in planetary studies.

THE HORN REFLECTOR AT BELL TELEPHONE LABORATORIES, USA

It is indeed remarkable that Bell Telephone Laboratories, (BTL), through their research interests in radio communication, have made such fundamental contributions to radio astronomy. One of these, the measurement of celestial background radiation was accomplished with a rather unusual type of aerial, which must be classed as a radio telescope because it provides the most accurate means of determining the absolute power flux received over an aperture. When in 1960–1 BTL participated in Project Echo, the experiment in long-distance communication via a reflecting sphere launched into orbit around the Earth, they employed a horn reflector of a type originated at BTL in the early 1940s. This combination of a horn and reflector is shown in Plate 4.18. Among several ideal characteristics, it is extremely broadband, has calculable aperture efficiency, and the back and sidelobes are so minimal that scarcely any thermal radiation is picked up from the ground. Consequently, it is an ideal radio telescope for accurate measurements of low levels of power flux. How these led to a highly significant determination of weak background radiation is related in a later chapter (p. 173).

Plate 4.18 The reflecting horn at Bell Telephone Laboratories, USA (*by courtesy of Bell Laboratories, USA*)

MASERS AND PARAMETRIC AMPLIFIERS

The realisation of ultimate sensitivity in radio astronomy depends not only on designing large radio telescopes but also on the minimising of receiver noise, predominantly the noise from the input stage which undergoes the full amplification of the receiver. At long wavelengths, of course, the noise from the Galaxy presents a limiting factor, but at wavelengths of several cm, receiver noise has in the past been a serious limitation to the detection of weak radio emission. The application of masers and parametric amplifiers in the late 1950s radically improved receiver sensitivity. The maser (standing for Microwave Amplification of Stimulated Emission of Radiation) depends on the increase in the population of selected atomic energy levels by means of a 'pumping' frequency, so that the amplification of the radio frequency becomes possible through the release of stimulated emission. After Townes and his colleagues at Columbia University in 1955 had demonstrated an ammonia maser oscillator, Bloembergen (1956) proposed a 3-level solid state maser which could be adapted as a very low noise preamplifier. The first application in radio astronomy was made by Giordmaine, Alsop, Mayer and Townes (1959) with a maser at 3 cm wavelength placed at the focus of the 50 ft reflector at NRL. It has been found practicable to reduce the effective noise temperature of masers, cooled with liquid helium, to no more than a few degrees absolute, most of which is contributed by losses in the input waveguide.

Although parametric amplifiers, which depend on a variable reactance induced in a solid-state or other device by a pump oscillator, cannot be reduced to such low effective noise temperatures, they constitute simpler, more convenient practical devices. Cooled by liquid nitrogen, effective temperatures can be brought down to about 20 K. Since there are normally present extraneous sources of noise that are difficult to eradicate, the parametric amplifier is usually adequate and the more popular device. Nevertheless, the maser must be resorted to if optimum sensitivity is essential, as in the Bell Telephone measurements of celestial background. The output sensitivity depends, of course, on the input bandwidth and the output integration time. Thus, with a 10 MHz input bandwidth and

10 seconds integration the resultant sensitivity is reduced to a ten thousandth part of the noise temperature of the receiver.

OTHER NOTABLE RADIO TELESCOPES AND OBSERVATORIES

Almost every country in the world is involved in radio astronomy research. There are about as many major radio astronomical stations as there are optical observatories. Japan,

Plate 4.19 The 100 m radio telescope near Bonn, West Germany (*by courtesy of the Max Planck Institut für Radioastronomie*)

for instance, is noted for its fine solar radio observatories. A very large Mills Cross is in operation at Bologna, Italy. The 100 m parabolic reflector completed in Germany in 1970, shown in Plate 4.19, is at this time the largest fully-steerable precision radio telescope. As I have already indicated, it would require a lengthy treatise to describe the variety and quality of radio telescopes throughout the world. Rather than attempt an exhaustive survey, I have covered many salient features; my account is intended to convey an impression of the types of radio telescopes and the problems encountered, Nevertheless, I conclude the chapter with some trepidation, because I am conscious that many notable radio observatories have not even been mentioned. I can only plead this is no reflection on their status and work. If radio astronomers in some distant land feel I have not allocated them due recognition, at least they are in good company, for I have said nothing of such renowned centres as Harvard, Illinois, Maryland, Stanford, to mention a few in USA alone. I can only express my apologies, and crave their indulgence, for I have already said enough to illustrate the scale and variety of radio astronomy installations. I must now proceed, in the next two chapters, to describe some of the astronomical achievements of these magnificent radio telescopes, and to endeavour to summarise two decades of research progress since 1952.

CHAPTER 5

The Solar System

THE RADIO SUN

In radio astronomy there are so many competing research topics that specialisation is inevitable; in consequence, solar radio investigations have tended to become the prerogative of certain observatories. After the first fashionable phase in the immediate post-war years when most radio astronomers were observing the Sun, many were deterred from further pursuit by the very complexity of solar activity; consequently, the pressing problems of the discrete sources and studies of the Galaxy tended to be accorded dominant priority. The complications of solar activity necessitated the development of specific techniques solely for radio observation of the Sun. The concentrated efforts of several research teams have been richly rewarded as a coherent picture of radio phenomena began to emerge. There are strong arguments in favour of intensive solar research. The Sun presents a unique astrophysical laboratory, close enough to observe in detail a great variety of high temperature magnetoactive plasma phenomena relevant to many processes which occur elsewhere in the universe. The real difficulty is in specifying exactly the circumstances surrounding any particular solar event. Nevertheless, with advances in detailed analysis there is great potential in solar research. It is undoubtedly a field where, in some aspects, radio methods are supreme, for although the total radio energy may not be large it is readily observable. And total solar flux at radio frequencies may increase up to 10^5 times during solar activity, while the changes in total optical power never exceed 1 per cent. Radio brightness temperatures in active regions can attain as much as 10^{11} K. Also, a potent means of studying the far out regions of the solar corona has been opened up by examining its influence on the propagation of radio waves. There are at least two excellent books on solar radio astronomy, by Kundu (1965), and by Zheleznyakov (1970) and I shall here describe only a selection of the results that have emerged from radio studies of the Sun; these can be grouped under four main headings, the quiet Sun, the radio plages, solar bursts, and the outer corona.

THE QUIET SUN

The quiet Sun is best observed, of course, during the minimum of the sunspot cycle, with appropriate subtractions or extrapolations to allow for any residual sunspot activity. I have already mentioned in Chapter 2 the initial observational and theoretical work on thermal radio emission from the solar atmosphere. Subsequent investigations brought further refinements, particularly in the two-dimensional patterns of brightness distribution derived from solar eclipses and by narrow-beam telescopes. A striking example of the latter was the radio map of the quiet Sun shown in Figure 5.1, obtained by Chris-

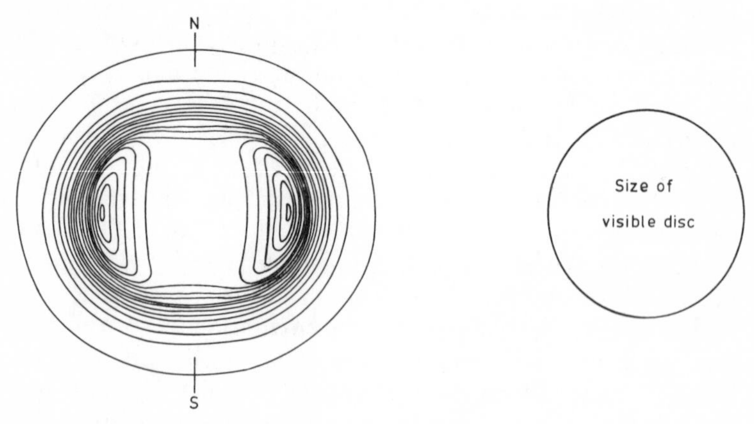

Figure 5.1 Radio contour map of the quiet Sun at $\lambda \approx 21$ cm (*After Christiansen and Warburton, 1953*).

tiansen and Warburton (1953) with the aid of grating interferometers. Radio and optical information about conditions in the solar chromosphere and corona have been in substantial agreement. The data are complementary, leading to more certain knowledge of electron density and temperature.

RADIO PLAGES

The association between sunspots and the slowly-varying or S component of radio emission at wavelengths of the order of 10 cm was first established by Covington (1948–9) who also noted that the radiation is partially circularly polarised. It was afterwards realised that the S component is more closely correlated both in size and duration with the chromospheric plages, the bright longer-lasting regions visible in the optical lines of hydrogen and calcium surrounding and overlying the sunspots. The radio sources, which may be called radio plages, were found to possess brightness temperatures of about 10^6 K and to move faster than the optical plages across the solar disk, indicating a greater height in the solar atmosphere. All these properties of the S component were in good accord with the interpretation that the radiation represented thermal emission from coronal condensations above the optical plages, regions of high density in the corona which had been recognised by Waldmeier and Müller (1950). The only anomaly, the apparent departure from the thermal spectrum at short wavelengths, has been explained by Ginzburg and Zheleznyakov (1959, 1961), as the effect of resonance absorption in the magnetic field (which must be present to account for the partial circular polarisation). In any case, the source is unlikely to be homogeneous; for instance, Kundu (1959) at $\lambda = 3$ cm, and Salomonovich (1962) at $\lambda = 8$ mm distinguished a bright nucleus surrounded by a diffuse area of emission. The plage regions are especially interesting because they have been found to be the main source of continuous X-ray emission from the Sun and have a major controlling influence on ionospheric ionisation, as well as being the areas where solar flares are manifested.

SOLAR RADIO BURSTS

Solar bursts, which frequently accompany solar flares, are comparatively simple in character at the short wavelengths, and the original classification by Covington of NRC Canada was based purely on amplitude changes. The microwave bursts have also been studied in other countries, particularly in USA, France, Germany, Russia and Japan, and it is generally agreed that the bursts can be divided into three main types; (a) an impulsive burst of 1 to 5 minutes duration closely associated with the visible flare, (b) a relatively weak and gradual rise and fall which may start before and continue after the flare, (c) a strong increase of intensity following the maximum phase of certain large flares and of duration normally exceeding 5 minutes. The impulsive and gradual bursts (a), (b) have been found to be closely associated with enhanced X-ray emission, measured with the aid of balloons and rockets. The separate significance of bursts of type (c) was recognised by Kundu and Haddock (1961) who called them microwave outbursts. Large outbursts are often the precursor of the arrival of high-energy solar protons in the Earth's atmosphere ten minutes to an hour or so afterwards. The microwave outburst often extends to longer wavelengths where it had been known as Type IV, and in 1957 Boischot and Denisse had recognised the association of type IV bursts with solar cosmic rays. A possible explanation of the connection between the ejection of solar cosmic rays and the microwave or Type IV outbursts has followed from the theory proposed by Takakura (1962) that the radio emission is synchrotron radiation from electrons which have been trapped and accelerated to relativistic speeds in the magnetic field of the active region. It is believed that the solar cosmic rays are the relativistic protons escaping from these regions.

It is interesting to recall that Forbush in USA first recorded an increase of cosmic ray intensity connected with a bright solar flare on 28th February, 1942. By a remarkable fortuity this striking event coincided with my recognition of the intense solar emission which occurred on 27, 28 February, 1942. Thus, two profoundly important discoveries were made independently at the same time; and twenty years had to elapse before a detailed relation between the two kinds of phenomena could be established.

Solar bursts at metre wavelengths are stronger, more complex phenomena, and Wild's group in Australia deserves the main kudos for disentangling the different types. The principal parameters which led to distinctions between the various kinds of bursts were the temporal variation of amplitude and spectrum, the location and movement of the source, and the polarisation. Wild and McCready (1950) differentiated three types, which they designated I, II, III, on the basis of the dynamic spectra obtained with their swept-frequency receiver. Bursts at metre wavelengths are now divided into five main types, and I will briefly describe each in turn.

Type I bursts appear as a multitude of sharp spikes of duration between 0·1 and 10 seconds, superimposed on the enhanced radiation which accompanies many sunspots. These noise storms which may last hours or days are a prominent feature of metrewave activity during sunspot maxima. The storm radiation, including the Type I bursts, was found to be strongly circularly polarised. The high radio brightness temperatures, up to 10^{11} K, clearly signify a non-thermal process, although the precise mechanism remains uncertain.

Type II bursts are the striking metre-wave outbursts of intense radio emission of several minutes duration occurring in conjunction with many solar flares, especially the very bright flares. The spectrographic observations by Wild (1950) showed a drift from higher to lower frequencies, and the sweeping interferometer measurements of position by Payne-Scott and Little (1952) indicated the outward movement of the source of radiation. Subsequent investigations, for instance by Roberts (1959), confirmed these characteristics. The frequency drift, at about 1 MHz per second, and the movements in position, were consistent with the interpretation that a stream of particles, presumably a shock wave, excited radiation at the plasma frequency as it moved out through the corona. The speeds deduced were similar in magnitude to the speeds of the corpuscular streams which produce magnetic storms on the Earth a day or two after large flares. Wild, Murray and Rowe (1953, 4) demonstrated the existence of a second harmonic in the radio bursts, but positional measurements by Smerd, Wild and Sheridan (1962) brought a curious result—sources of harmonics apparently generated backwards into the solar atmosphere. Fortunately, a theoretical paper by Ginzburg and Zheleznyakov (1959) rescued the situation from its seeming difficulties of interpretation, as well as explaining how the influence of plasma waves on coronal irregularities excited radio emission.

Type III bursts, often occurring at the commencement of a flare, are short-duration bursts showing a rapid frequency drift, about 20 MHz per sec. The diagram, Figure 5.2, obtained

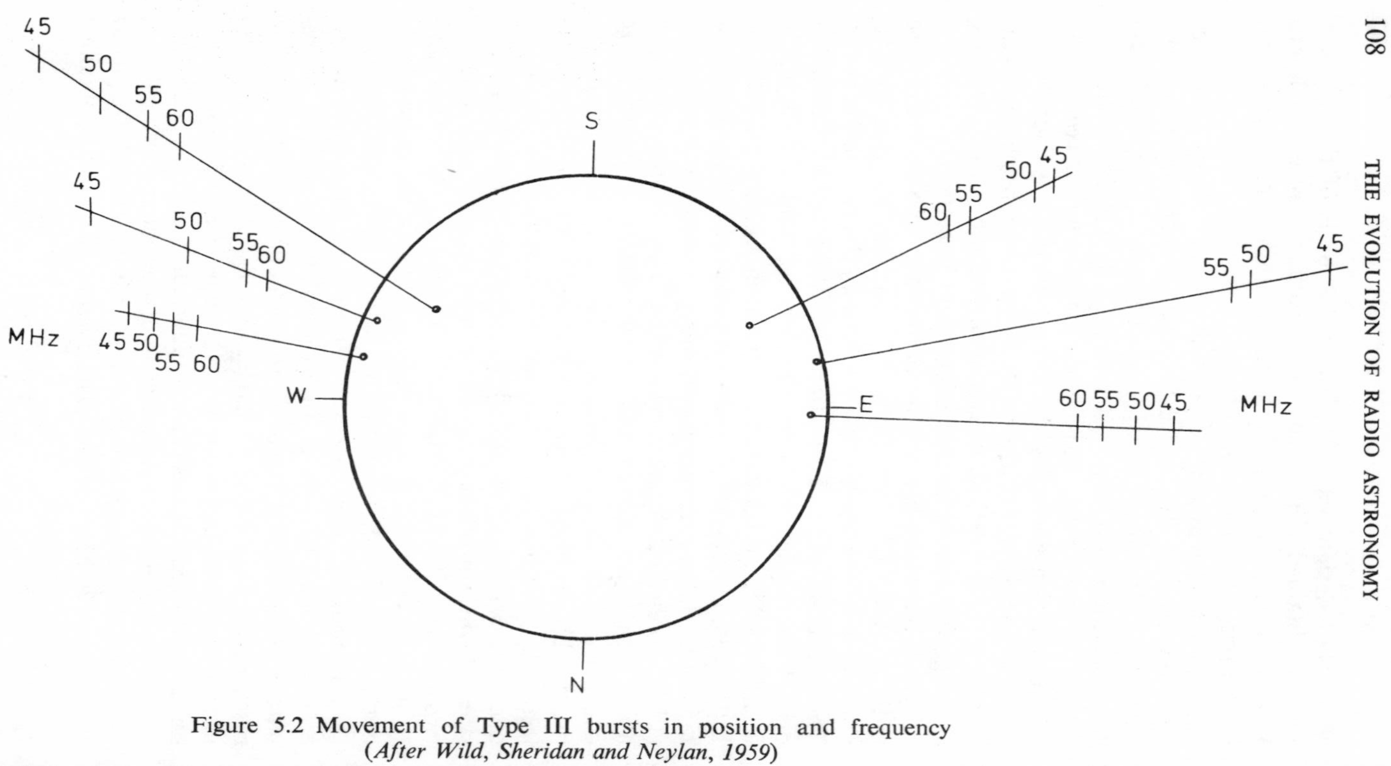

Figure 5.2 Movement of Type III bursts in position and frequency (*After Wild, Sheridan and Neylan, 1959*)

by Wild, Sheridan and Neylan (1959) with a swept-frequency interferometer, strikingly demonstrates the outward movement of the radio source. As in Type II, the radiation is unpolarised and there is often a second harmonic. The fast outward movement has been interpreted as the excitation of plasma wave radiation by very fast puffs of electrons ejected with speeds about $\frac{1}{4}$ that of light.

I have already mentioned the Type IV burst, a long duration burst covering a very wide frequency band, and which sometimes follows Type II. The characteristics of the Type IV burst were first established by Boischot in 1958 with the Nancay interferometer in France and he suggested that the emission is produced by the synchrotron process.

Type V is a broad band radiation lasting for a minute or so, and was first recognised by Wild, Sheridan, and Neylan (1959) as a distinct type which sometimes follows Type II bursts.

The diagram (Figure 5.3) illustrates the dynamic spectra of

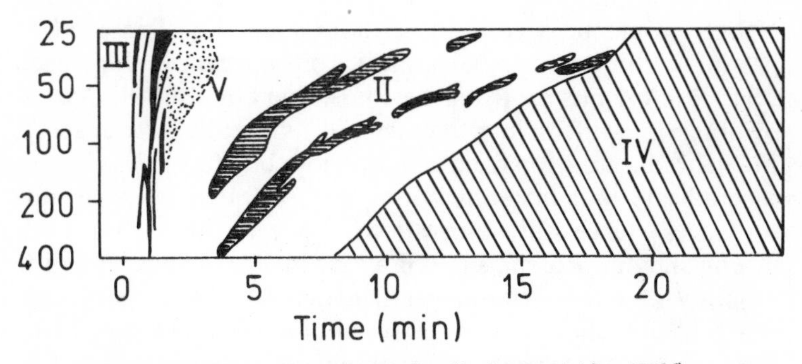

Figure 5.3 Dynamic spectra of radio bursts (*After Wild*).

the different types of metre-wave bursts that may occur during a large solar outburst. As well as the main types, there are many sub-varieties that sometimes appear, for instance the U bursts found by Maxwell and Swarup (1958) where a Type III burst starts with a normal frequency drift and then reverses. Although it is abundantly evident that the whole subject of radio bursts is extremely complex, it is fascinating and revealing. Radio observations are providing unique information about many types of coronal activity and eruptive motions. With the extension of studies of the Sun to cover simultaneously the whole radio band from millimetres to metres we can anticipate that further great advances will be made. Two years of observations with the Culgoora radioheliograph reviewed in a paper by Wild (1970) indicates a wealth of new information. As more detailed data in the portrayal of solar activity becomes available we may be certain that what may seem at first sight to add further complications will ultimately lead to a great clarification of our understanding of solar atmospheric physics and the nature of solar phenomena.

THE OUTER CORONA AND THE SOLAR WIND

The possibility of studying the solar corona by observing the refraction of radio waves at metre wavelengths from distant sources whose directions lie close to the Sun was suggested independently by Vitkevich in Russia and by Machin and Smith in Cambridge in 1951. They realised that the Crab Nebula would be a convenient radio source, its line of sight passing within about 5 solar radii from the Sun in June each year. Radio emission from the whole Sun could be eliminated by observing with an interferometer of sufficiently large spacing. Intense radio emission from active regions on the Sun was harder to eradicate and first attempts were frustrated by solar disturbances. Successful observations, made in June 1952 both in England and Russia, showed a very marked reduction in the visibility of the interferometer fringes from the Crab Nebula when its direction passed close to the Sun. The surprising feature was the decrease in amplitude at distances as great as $10R_{\odot}$, which could not be explained by simple refraction. To account for the reduction in fringe visibility it was postulated that scattering by irregularities in coronal electron density must produce an increase in the apparent diameter of the radio source. The interpretation was amply confirmed in the subsequent experimental studies and theoretical analyses expounded by Hewish and by Vitkevich in 1955. Further investigations at Cambridge (Hewish, 1958, Högbom, 1960) and Russia (Vitkevich, 1960) revealed that the electron irregularities were elongated in directions which corresponded well with the filaments of the radial coronal streamers optically visible during solar eclipses. By now, scattering had been detected out to $30R_{\odot}$, and showed evidence of an increase at the maximum of the sunspot cycle. A significant extension in the study of the far-out corona was achieved by Slee (1961) who, using a long interferometer baseline of 10 km separation with a radio link, observed the passage of a total of 13 radio sources at $\lambda = 3 \cdot 5$ m along a variety of tracks relative to the Sun. In this way, he succeeded in measuring small amounts of scattering and covered a much greater extent of the corona. The scattering was detected furthest in the equatorial directions—out to a distance of about $100R_{\odot}$

Meanwhile, the relation between the interplanetary medium and the solar corona had been receiving the attention of theoretical physicists, and Parker (1960) had concluded that the solution of the hydrodynamic equations for the solar atmosphere predicts a large-scale expansion at supersonic velocities. Space probe measurements lent support to Parker's theory of the solar wind, indicating a particle flux density of about 10 per cm^3 and a velocity about 300 km/sec in the vicinity of the Earth. Following continued studies of radio scattering at $\lambda \sim 8$ m, Hewish and Wyndham (1963) concluded that their results were substantially in agreement with Parker's model.

The next advance came as a surprise, although as so often happens, with hindsight it could be argued that it might have been foreseen. Radio scattering from an irregular distribution of electron density can manifest itself in several ways. The inhomogeneities affect the phase of the waves so that the emergent wavefront is distorted. In consequence the scatter of apparent directions spreads the observed angular extent of the source. Also, the received amplitude may be altered since the waves from different parts of the front no longer hold the same phase relation. Consequently, moving irregularities cause fluctuations of received intensity, but these are averaged out if the source is large. What happens in particular circumstances depends on the scale of the irregularities and the size of the source.

By chance, M. E. Clarke, during a Cambridge survey in 1963, noticed that certain sources of small angular size exhibited continual rapid scintillations which could not be attributed to the ionosphere. In a further investigation, Hewish and his colleagues realised that the scintillations were produced by the interplanetary medium, and that the fluctuations would only be apparent on sources less that 1″ in angular size. I shall discuss later, in Chapter 7, how the discovery led to a valuable method for estimating the sizes of very small sources. Actually, the possibility that variations in ionisation of the interplanetary medium might produce scintillations had been suggested by Ginzburg in 1956. From the point of view of the present discussion on the Sun the importance of the observations lies in the information they yield about the parameters of the solar wind. By recording on spaced receivers at the corners of a triangle of sides about 50 to 80 km, Hewish and Dennison (1967) deduced that at a distance of 0·5 AU from the Sun the speed of the solar wind was about 300 km/sec and the dimensions of the irregularities of the order of 200 km. In this interesting field, the study of the interplanetary medium far from the Sun, the radio measurements have entered a province now partially explored by the recording probes of space vehicles.

RADIO WAVES FROM THE MOON AND PLANETS

In general terms, we may regard the Moon and planets as inert bodies whose temperatures are mainly determined by the amount of heat they receive from the Sun. Computed temperatures, ranging from a few hundred degrees absolute for the Earth and inner planets, down to about 50 K for distant planets like Uranus, conform reasonably well with infrared measurements. Bearing in mind the Rayleigh-Jeans law, power flux proportional to T/λ^2, it was obvious that radio observations of planets, with their low temperatures and small angular size, would be a severe test of radiometric sensitivity. Nevertheless, radio offered the prospect of providing valuable information. Observing over a wide range of wavelengths would present the opportunity of deciding gross surface characteristics, and of penetrating optically opaque planetary atmospheres like that of Venus. By appropriate choice of wavelength, radio had the ability to probe into dense atmospheres, and to assess subsurface temperatures since radio waves penetrate the solid material to a depth of the order of the wavelength. The hardest task is to attempt to deduce detailed surface composition and structure, a difficulty soon apparent in the endeavour to decipher the data of radio emission from the Moon, which I shall now discuss.

RADIO EMISSION FROM THE MOON

Compared with the planets, the Moon is a comparatively simple object to study, with its large angular size and absence of any appreciable atmosphere. I have previously mentioned (page 47) the measurements of radio brightness temperature throughout the lunar cycle made by Piddington and Minnett at $\lambda = 1 \cdot 25$ cm in 1949. They attempted a comprehensive analysis of the data based on the theoretical equations of equilibrium between the radiation at the surface and heat flow through the lunar material. It was soon realised that the lunar surface was composed of a substance with low thermal capacity and conductivity, probably consisting of dry, porous rock like pumice or in a granular form. It was evident that the more precise solution to the problem was wrapped up in a profusion of parameters. These include the density, specific heat, thermal conductivity, dielectric constant, roughness and topography of the surface, whether the sub-surface is homogeneous or in layers, and whether there is any interior source of heat in the Moon. In such circumstances it was clearly desirable to make measurements over as wide a range of circumstances as possible, for instance, by observing at different wavelengths, during lunar eclipses in addition to the normal lunar phases, and to improve resolution in order to examine the distribution of emission over the lunar surface. By suggesting likely values of the various parameters, it might then be possible to find a model which would fit the observations. Gradually, more data was accumulated; at NRL for instance, Gibson (1958) determined the cyclic phase variation for the centre portion of the disk at $\lambda = 8$ mm, and demonstrated that the radio emission was unaffected by the rapid change of illumination during a lunar eclipse. Coates (1961) at $\lambda = 4$ mm found that during the lunar cycle the maria, with the exception of Mare Imbrium, heated up and cooled more rapidly than the surrounding mountainous regions—adding a new complication to interpretation. In Russia, Soboleva (1963) observing at $\lambda = 3$ cm with the 1′ fan beam of the Pulkova multi-plate radio telescope, successfully demonstrated how the distribution of radio emission depends on polarisation as forecast by Troitsky (1954), since the reflection coefficient and hence the emissivity of an inclined surface depends on polarisation. The 22 m radio telescope at Pulkova was used by Salomonovich (1962) to map the emission over the lunar surface at several wavelengths between 8 mm and 10 cm. Comprehensive analyses of available data were made by Troitsky (1965) and others. It will suffice here to say that the research has been an instructive scientific exercise although it was far from easy to reach unambiguous detailed conclusions much beyond those derived from the early observations. After some initial debate, it was generally agreed that it was unlikely that lunar explorers would be harassed by more than a thin layer of surface dust. With the advent of lunar landings, the subject has become academic rather than of any practical concern, except in so far as the same type of problem must arise with respect to the planets. There is, of course, a complementary contribution from radar astronomy which I shall discuss later.

RADIATION FROM JUPITER

Jupiter was the first planet to be observed by radio. Once again, there occurred a dramatic, totally unexpected radio astronomical discovery, originally ascribed to "interference". It is an exhilarating reward for a scientist during his research to find an unforeseen phenomenon breaking through by its impact on his observations. Even so, but for the persistence and perspicacity of two observers, Burke and Franklin, the discovery might easily have been missed. Certainly, no one imagined that intense radio bursts could arise from Jupiter. The story of the discovery of this amazing radiation in 1955 was the culmination of an interesting sequence of events.

When Burke was a research graduate at MIT his interest in radio astronomy was awakened through a lecture given by Bowen, while on a visit from Australia. After Burke had completed his Ph.D. thesis, he seized an opportunity to move to the Carnegie Institute where Graham Smith, on sabbatical leave from Cambridge, had set up an interferometer to study radio source scintillations. In June, 1954, when the direction of the Sun came close to the Crab Nebula, they decided to examine at 22 MHz the influence of the solar corona on the radiation received from the Crab radio source. Occasionally interfering bursts were noticed which they attributed to solar activity. Their next project was to make a survey of radio sources at 22 MHz, for which purpose a Mills Cross was designed by Smith and built by Burke. By the time the Cross was completed early in 1955, Franklin, an astronomer, was working in collaboration with Burke. During their survey, the "interference" bursts appeared again in a direction not very far away from the Crab Nebula although by now the Sun was no longer in the vicinity. Could ignition interference from a passing vehicle be responsible, they pondered. Then Burke perceived that the bursts always occurred at about the same sidereal time, showing that they must originate from a celestial source. However, measurements over a three month interval revealed a gradual drift. H. Tatel, who had been examining Jupiter optically, facetiously remarked to them that Jupiter might be the source. When Burke and Franklin completed their radio recordings that evening the sky was fine and clear with Jupiter shining brightly in the specified direction, yet the idea of a possible connection seemed too far fetched. Next day, however, Franklin accurately plotted the direction of Jupiter over the period of their observations, finding a precise correspondence which led to only one conclusion, that the intermittent bursts must emanate from Jupiter. The bursts are so strong at about 20 MHz that only solar bursts exceed their intensity. What could be their origin? One plausible conjecture, thunderstorms in the turbulent Jovian atmosphere had to be abandoned, because the spectrum was so different and the bursts too intense to explain them in this way.

News of the discovery led Shain in Australia to re-examine the recordings he had made in 1951 of cosmic radio noise at 18 MHz. Sure enough, the Jupiter bursts were there, Shain having previously disregarded them as interference. Nevertheless, Shain (1956) now made a very fruitful analysis of the recordings, and he found a periodic variation in the bursts, with a marked peak at each planetary rotation. Despite many subsequent studies of the detailed properties of the decametric Jupiter bursts, it has remained difficult to reach any decisive conclusion as to how they are produced. Amongst their significant characteristics they show pronounced circular polarisation indicating the presence of a fairly strong magnetic field. Interferometer measurements of the size of the emitting region by Slee and Higgins (1964) showed that the source occupied no more than about 1/30 of the size of the Jupiter disk. One of the most curious results of all, found by Bigg in 1964, was that the probability of occurrence of bursts was strongly related to certain orbital positions of Io, the closest of the large satellites of Jupiter. Evidently, Io was causing a tidal or magnetic disturbance stimulating radiation from the Jovian ionosphere or magnetosphere.

Unexpected radio properties also appeared at shorter wavelengths. Detection of Jupiter at cm wavelengths was first achieved in 1958 by Mayer and his colleagues at NRL, Washington, with their 50 ft radio telescope. Their measurement of an effective temperature of about 145 K at 3 cm wavelength indicated that the radiation at this wavelength must be predominantly thermal. At 10 cm wavelength, however, Sloanaker (1959) recorded a temperature of 600 K, while Drake and Hvatum (1959) found that at 70 cm it had risen to 50 000 K. It was obvious that this steady emission at decimetric wavelengths, 10 cm and above, must possess a strong non-thermal component. Drake and Hvatum suggested that it might be attributable to synchrotron radiation from magnetically trapped electrons surrounding Jupiter. It seemed likely that such a large planet with a strong magnetic field might give rise to more intense regions than the van Allen belts which surround the Earth. The hypothesis was fully vindicated in 1960 when Radhakrishnan and Roberts demonstrated with the aid of the Cal. Tech. interferometer, that the equatorial radio diameter was about 3 times that of the planet and contained a strong linearly polarised component. At the same radio observatory, Morris and Berge (1962) discerned a periodic rocking in the direction of polarisation, from which they inferred that the magnetic axis must be tilted at 9° with respect to the axis of rotation. Jupiter is the only planet for which there is definite evidence of non-thermal radio emission.

THE ATMOSPHERE OF VENUS

If the characteristics of radio emission from other planets seem tame compared with the profusion of Jovian phenomena, the radiation from the atmosphere of Venus has been particularly interesting and revealing. The dense white atmosphere obscures the surface from optical view, and the infrared radiation from the cloud cover gives a temperature of 225 K in good agreement with the planetary temperature expected from solar heating. Venus was the first planet to be detected at centimetric wavelengths, by Mayer and his colleagues (1958) using the NRL 50 ft reflector, and they were surprised to find the radio brightness temperature at wavelengths of 3 cm and 10 cm to be about 600 K. The intensity of the radio waves emerging through the dense cloud coverage provided for the first time direct evidence of the very high temperature at the surface of Venus. At 8 mm wavelength, Gibson and McEwan (1959) measured a temperature of 400 K, showing that some of the radiation at this wavelength came from cooler parts of the atmosphere. High resolution observations at 10 cm made with the Cal. Tech. interferometer by Clark and Kuzmin (1965) demonstrated differential polarisation, showing that at least the main part of the 10 cm radiation originated from the solid surface. The surface heating has been explained as a 'greenhouse effect', mainly attributable to the large amounts of carbon dioxide in the atmosphere of Venus.

RADIO EMISSION FROM OTHER PLANETS

The radiation from Mars, with its comparatively clear and tenuous atmosphere, occasioned no surprises, and the temperature ($\sim$210 K) recorded by Mayer *et al.* (1958) was in accordance with expectations. Mercury presented a more difficult target for radio observation because of its small size and proximity to the Sun. It was detected by Howard, Barrett and Haddock (1962) at about 3 cm wavelength using the 85 ft Michigan radio telescope. They found a mean disk temperature at maximum elongation of about 400 K, a higher value than had been anticipated. It had been believed that Mercury always kept the same face to the Sun with a temperature of about 600 K, while the dark side was intensely cold. The radio measurements were puzzling, for they suggested a relatively warm sub-surface temperature on the dark side. The enigma was resolved a few years later when radar echoes from Mercury proved that the planet is slowly rotating so that the whole surface is periodically warmed by the Sun. By 1966, Saturn, Uranus and Neptune had all joined the list of planets detected by their radio emission. Although recorded temperatures at some wavelengths were rather higher than had been anticipated, the differences have appeared explicable as possibly due to atmospheric "greenhouse" effects or partly to internal heat sources within the planets. No planet other than Jupiter has exhibited extraordinarily intense radiation.

RADAR ASTRONOMY

Astronomy by radar has the great attribute of accurate determination of distance. Unfortunately, the maximum range is limited, firstly by the power attainable in manmade transmitters, secondly by the meagre fraction of incident flux reflected from the target, thirdly by the rapid diminution of echo strength with distance. These limitations have confined radar astronomy to the solar system, where impressive results have been obtained, as I shall now describe.

RADAR OBSERVATIONS OF THE MOON

The Moon echo experiments of the U.S. Army Signal Corps in 1946, and by Kerr and Shain (1951) utilising the short-wave broadcast station "Radio Australia", were hampered by slow, deep fading. The cause of the fading was explained by Murray and Hargreaves at Jodrell Bank in 1954; they showed that the magneto-ionic action of the ionosphere twisted the polarisation of the radio wave so that the echo often returned in a different plane of polarisation from that of the receiving aerial. It was one of those effects so obvious in retrospect, for Faraday rotation was well known in ionospheric radio at longer wavelengths. The whole difficulty was overcome, in what has now become standard procedure, by using circular polarisation for radar waves passing through the ionosphere.

A rapid fluctuation of echo strength was also present, which

Kerr and Shain correctly attributed to the apparent slight oscillation of the Moon known as libration and the consequent beating together of echoes from different parts of the lunar surface. Obviously this type of fading would depend on the roughness and topography of the Moon, and the first systematic study of radar scattering was made at Jodrell Bank by Browne *et al.* (1956) and Evans (1957). All the early experimenters had worked with radar pulses longer than the 11·6 milliseconds required to envelop the whole lunar hemisphere. Later investigators used short pulses with great advantage, showing how the echo altered with inclination to the surface. Figure 5.4 shows results obtained in 1963 by Evans and Petten-

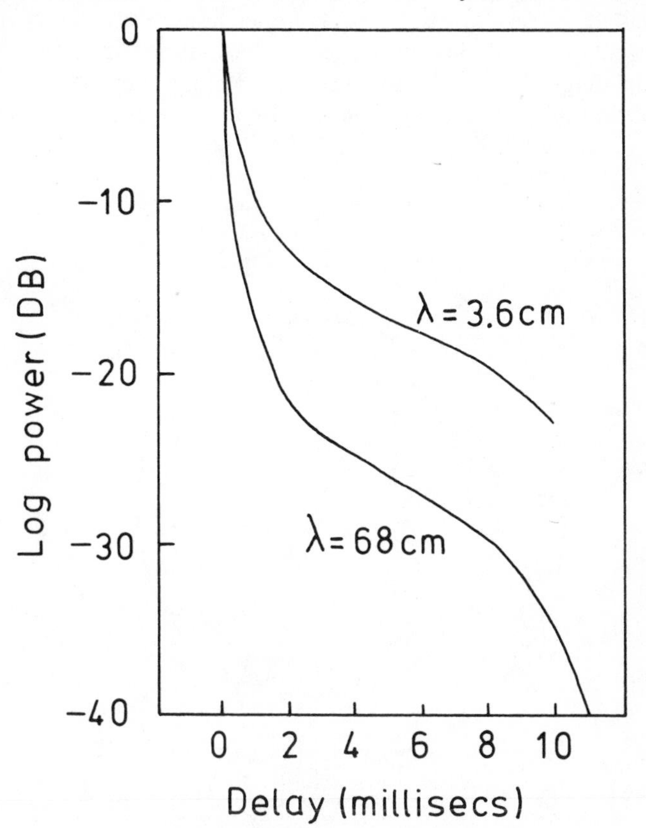

Figure 5.4 Dependence of echo power on distance from nearest point of the Moon (*After Evans and Pettengill, 1963*).

gill (Evans had by then moved from Jodrell Bank to Lincoln Lab., MIT), and the greater scattering towards the limb of the Moon at the shorter wavelength is indicative of small scale roughness. By such investigations, average slopes were deduced to be about 1 in 7 on the metre-wave scale and about 1 in 3 on the centimetre scale.

The most spectacular advance in radar mapping of the Moon was achieved with a technique developed by Pettengill (1960). The surface of the Moon can be divided into circular range zones, and into strips of different Doppler shifts as illustrated in Figure 5.5, where A_1 and A_2 are regions corresponding to a particular range and Doppler shift. By analysing the reflected signal in terms of range and Doppler shift, the echoes from different parts of the surface could be mapped in detail. The radar map of the Tycho crater in Figure 5.6 obtained by Pettengill and Thompson (1968) is a remarkable demonstration

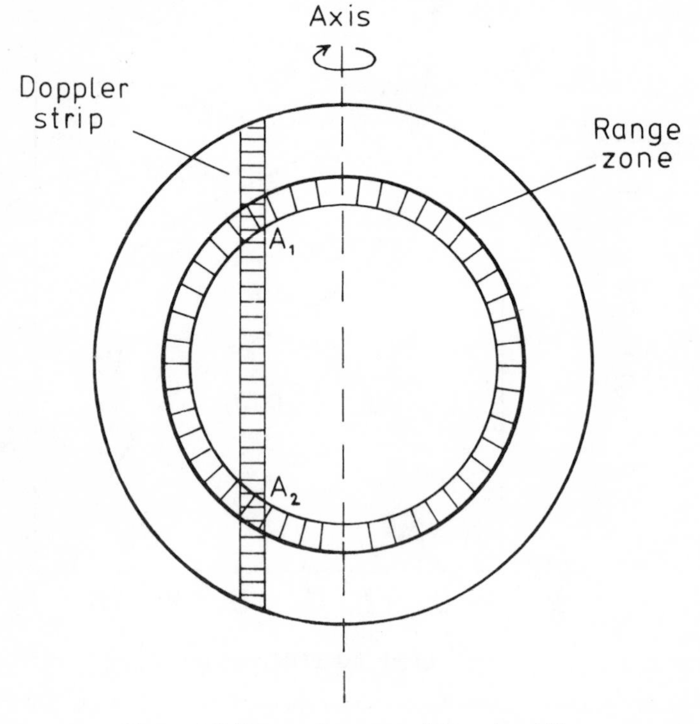

Figure 5.5 Zones of range and Doppler shift.

of the discrimination attainable by this technique. Another notable radar contribution was the very accurate measurement of the distance of the Moon made by Yaplee and his colleagues at NRL. By the 1960s USA had taken a strong lead in lunar and planetary radar observations.

The advances in lunar radar techniques could clearly be a great asset when transferred to studies of the planets. For instance, radar pulses could penetrate through the opaque atmosphere of Venus to inspect the surface and its topography. To attain the required sensitivity to observe the planets presented the most challenging problem. Although Venus comes

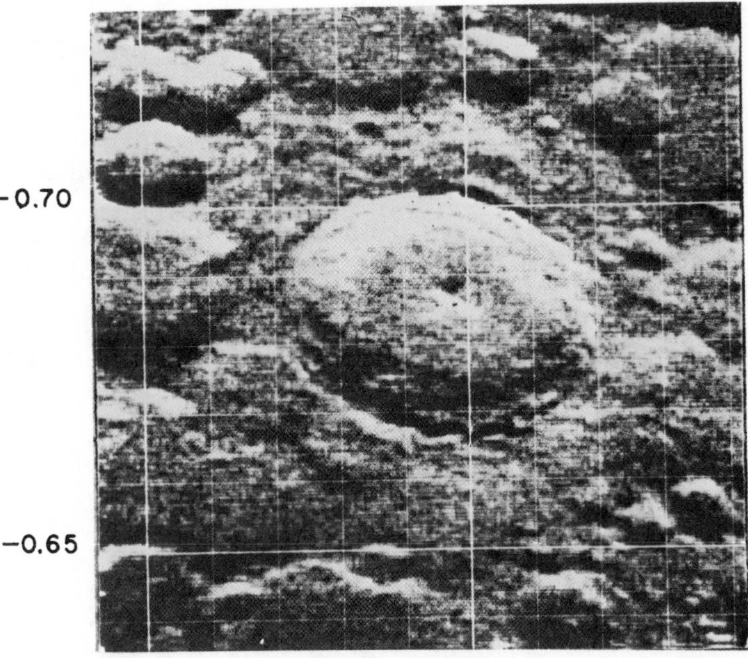

Figure 5.6 Radar map of the Tycho crater and surrounding region observed with Haystack radar at $\lambda = 3 \cdot 8$ cm (*After Pettengill and Thompson 1968*).

closer to us than any other planet, its great distance made it 10 million times harder to detect than the Moon. Radar technology took almost 15 years to overcome this factor. The first and most pressing task was to measure the distance to Venus, since an accurate value would decide the scale of the whole solar system, an essential yardstick both for space exploration and astronomy. Lincoln Lab., MIT, pioneered the attempts to detect Venus by radar. As the echo interval at nearest approach is about 5 minutes, the method employed was to transmit a series of pulses for 5 minutes, and in the next 5 minutes to record the echoes on magnetic tape. The digitised recordings were subsequently analysed and integrated with the aid of computers. The success first claimed in 1958 proved premature. Curiously enough, in the following year, Jodrell Bank confirmed the MIT range measurement, while MIT themselves were failing in a repeat experiment to observe echoes. As it turned out, both groups had interpreted spurious signals, only a few times the mean amplitude of random variations, as genuine, and their agreement in range was an extraordinary coincidence.

At the time of the nearest approach of Venus in 1961 a fresh onslaught was made with higher sensitivity to render more certain the validity of the results. On this occasion, five groups had assembled equipment: Lincoln Lab., MIT, and Jodrell Bank, and in addition, the Jet Propulsion Lab. (JPL), the Radio Corporation of America (RCA), and a group in Russia. All five were in close agreement, as illustrated in Figure 5.7, differing substantially from the earlier optical and radar estimates. On the basis of the new measurements, the International Astronomical Union in 1964, adopted 149 600 000 km as the officially recognised value of the astronomical unit of distance (the mean radius of the Earth's orbit). Radar observations, penetrating the featureless cloud cover, also revealed the rotation rate of Venus, and Carpenter and Goldstein at JPL in 1964 established a retrograde rotation with a period of about 250 days.

Radar also made an interesting contribution in assigning the rate of rotation of Mercury. The first radar detection of Mercury was achieved by Kotelnikov *et al* in 1962 in Russia, but the surprise came in 1965 when Pettengill and Dyce, observing at

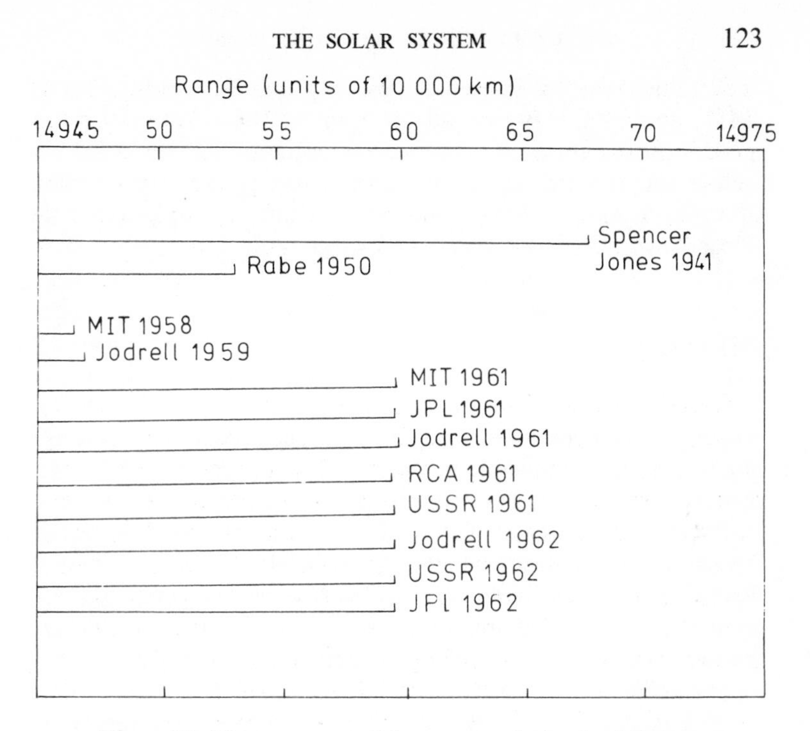

Figure 5.7 Measurements of the astronomical unit of distance.

430 MHz with the 1000 ft Arecibo reflector, deduced a rotational period of 59 days. Optical data obtained as long ago as 1889 were believed to have convincingly established that the planet always turned the same face to the Sun and so had a rotation period of 88 days equal to its orbital period. On reanalysis, it was realised that the optical data had been misinterpreted.

With the correct rotation derived by radar, it was now possible to account for the warm temperature of the dark side measured by radio emission since the rotation ensured periodic heating of the whole surface.

Planetary radar astronomy accomplished further advances, stimulated and encouraged by the interests of space exploration. Particularly notable achievements were made with the aid of the large steerable radar telescopes in USA such as JPL's Goldstone 210 ft reflector, and the 120 ft Haystack reflector at Lincoln Lab., both capable of operating down to cm wavelengths. By 1970 the radar mapping of Venus and Mars was well under way by Pettengill, Shapiro and their colleagues at MIT, and by Goldstein and his team at JPL. Planetary radar observations form a continuing programme designed to refine the determination of surface topography and orbital motion, thus providing valuable preliminary data prior to closer inspection with the aid of space vechicles.

METEORS

Radar methods have many applications in research on the upper atmosphere, which may be regarded as a subject belonging to the realm of geophysics rather than astronomy, and I will merely remark that many important contributions have been made, for instance, to the studies of aurorae and the incoherent scatter of electrons up to great heights in the ionosphere. Radar reflections from the ionised columns produced by meteors have yielded much valuable information about winds and physical conditions at heights between 80 and 120 km.

The radar study of meteor trails has also led to a vast amount of astronomical data on the orbits and other parameters of meteors. Radar methods, supplemented by improved optical techniques, revolutionised meteor astronomy, and by 1954 the time was ripe for Lovell's comprehensive treatise reassessing the whole subject. I will mention briefly only one or two salient points. The radar measurements of meteor velocities settled a long-standing controversy as to whether all meteors were confined to the solar system. Öpik had previously contended from visual data that many sporadic meteors had speeds exceeding the parabolic limit and must be of interstellar origin. The radar studies of meteor velocities by Almond, Davies and Lovell (1951, 2, 3,) and by McKinley (1951) proved beyond doubt that practically all meteors approached the Earth at speeds less than 72 km/sec and must have been orbiting the Sun. The orbits of sporadic meteors as well as meteor showers have been calculated from radar data. The perfection of radar techniques is typified by the three-stations method of Gill and Davies (1956). By simultaneously measuring range and velocity from three sites, complete information on radiants, velocities and orbits of individual meteors could be deduced.

RADAR ECHOES FROM THE SUN

One might have thought that the Sun would be the supreme radar target in the solar system, but its observation has presented special problems. The radar requirements for detecting echoes from the Sun were discussed by Kerr in 1952; it was evident that the highly-ionised solar atmosphere would reflect radio waves and that wavelengths around 10 m were advisable if undue absorption was to be avoided. Obviously, the radio noise emitted by the Sun would be a most irritating nuisance. The first radar echoes from the Sun were obtained in 1959 by Eshleman and co-workers at the Stanford Research Institute, USA, and in 1961 a more systematic study was begun by James and his colleagues of Lincoln Lab., MIT. Using an array of dipoles covering about 8 acres on a site in Texas, they transmitted 500 kW of power for about 17 minutes and received the echo during the next 17 minutes. The Sun is a complex target because of the turbulence and activity in the solar atmosphere. At the same time it is these features which make it such a potentially interesting target, for the measurements of Doppler shifts can indicate the velocities of mass motions and ejection of ionised streams. The outward flow of the solar wind has been detected. The initial results are reminiscent of the early research on solar radio emission, the complexities have so far tended to frustrate precise interpretation. Similarly, the detailed investigations which lie ahead will almost certainly be amply rewarded in the eventual outcome of knowledge of solar phenomena that is revealed.

CHAPTER 6

Radio Waves in the Galaxy

21 CM LINE STUDIES OF THE GALAXY

The realisation of the scheme envisaged by Oort and van de Hulst for the hydrogen line studies of the Galaxy is a supreme example of a well-planned research project. Their first step was to demonstrate theoretically and experimentally that the line, at $\lambda = 21$ cm, could be observed. The second stage was to determine the structure and kinematics of the Galaxy from measurements of the intensity and Doppler shifts of the line. It

Plate 6.1 H. C. van de Hulst at Leiden in 1968 (*by courtesy of H. C. van de Hulst, Leiden*)

was evident that the radio line could provide a means of deriving the mass and motion of neutral hydrogen throughout the Galaxy, with the advantage of being unaffected by the dust clouds which cause so much optical obscuration in the galactic plane.

The Leiden group, together with the CSIRO team at Sydney,

Plate 6.2 B. J. Bok and J. H. Oort (right) (*by courtesy of the Radiophysics Laboratory, CSIRO, Australia*)

were pre-eminent in the fulfilment of the plan. Muller was responsible for developing the Leiden radio equipment, and a comprehensive account of the methods of investigation was given in a paper by van de Hulst, Muller, and Oort in 1954. An extensive survey of the northern part of the Milky Way visible from Leiden culminated by 1957, after several years of assiduous observation and exacting analysis, in papers by Muller, Westerhout, and Schmidt. At CSIRO, a corresponding survey in the

Plate 6.3 A group of radio astronomers who took part in early studies of the 21 cm hydrogen line. Left to right: F. J. Kerr, J. P. Wild, J. V. Hindman, H. I. Ewen, C. A. Muller, W. N. Christiansen (*by courtesy of the Radiophysics Laboratory, CSIRO, Australia*)

southern hemisphere was completed in 1959 by Kerr, Hindman and Gum. The two sets of results were combined in the "Leiden-Sydney" map of the spatial distribution of neutral hydrogen in the Galaxy, Figure 6.1, which clearly shows the existence of spiral arms.

The precise placing of the arms on the map depended on correct assumptions being made about the rotation of the Galaxy. Knowledge of the relation between rotational velocity and distance from the galactic centre had to rely on optical data of stellar motions. Originally it was believed that a simple curve could be drawn representing circular velocity at different radial distances. The detailed fitting together of the northern and southern radio maps showed that although circular motion was valid as a first approximation, there were considerable local deviations, apparently due to differential streamings. By 1970

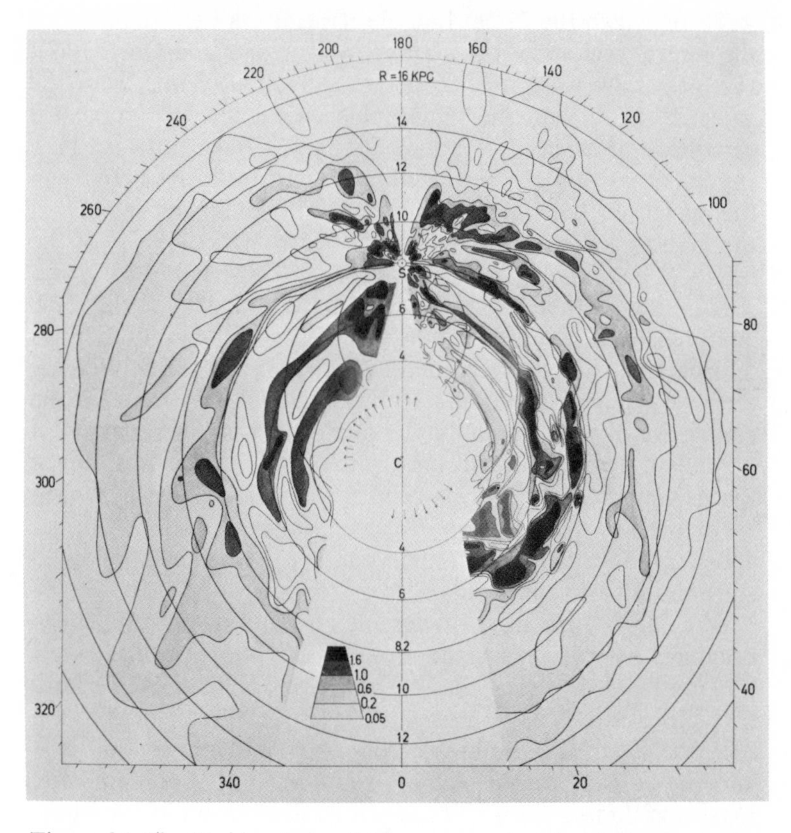

Figure 6.1 The "Leiden-Sydney" 21 cm map of neutral hydrogen in the Galaxy.

the interpretation of 21 cm line profiles had become a more uncertain task than it seemed 15 years earlier. This in no way detracts from the ultimate value of the method, but rather emphasises the need to clear up discrepancies before precise solutions can be claimed. Now, in the 1970s, a concerted attack is being launched to elucidate the galactic structure, by studies of the 21 cm line and the more recently detected radio lines of ionised hydrogen and various molecules, together with a renewed pursuit of optical data. We may confidently anticipate the emergence of a more finished and certain picture of the "grand design" of the Galaxy.

To return to the 21 cm line observations, an examination of the central regions of the Galaxy by Rougoor and Oort (1960) revealed a spiral arm at a distance of about 3 kpc from the galactic centre with the remarkable characteristic of outward expansion at a speed of about 50 km/sec. Such outward flow suggests that it might be the efflux of past explosive activity in the nucleus of the Galaxy. In contrast, in the far-out regions of the Galaxy there is evidence, described by Oort (1969), of the inflow of gas clouds attracted into the Galaxy from outer space.

No other method can trace out the entire Galaxy in the same comprehensive manner as the 21 cm line studies of neutral hydrogen have done. The method proved ideal for delineating the overall shape. The bulk of the hydrogen was found to be contained in a remarkably flat and thin disk, so making it possible to define more accurately than hitherto the position of the galactic equator and its centre. In this way, the hydrogen surveys led to the adoption in 1960 of a new system of galactic coordinates—a formal result, perhaps, but nevertheless a substantial basic contribution to astronomy.

Auxiliary information on the interstellar hydrogen has been obtained by measuring the 21 cm absorption profiles when looking towards strong discrete sources. The first line absorption observations were made in 1954 by Hagen and McClain with the 50 ft NRL radio telescope looking towards Taurus A and the galactic centre. Almost simultaneously, Williams and Davies at Jodrell Bank measured line absorption in the directions of Cygnus A and Cassiopeia A. The method permits the specific inspection of the hydrogen lying along the path to the source.

CONTINUUM RADIATION FROM THE GALAXY

Galactic radio emission, the original discovery in radio astronomy, deceptively simple at first scrutiny, has proved surprisingly difficult to disentangle. Despite the ease of detecting continuum radiation, and its obvious general correspondence with the Galaxy, certain aspects of the background emission have eluded attempts at complete clarification.

The contrast in the spectrum of thermal radio emission from ionised hydrogen, as compared with the non-thermal synchrotron radiation, has fortunately allowed these two types to be clearly distinguished. Typical spectra are illustrated in Figure 6.2.

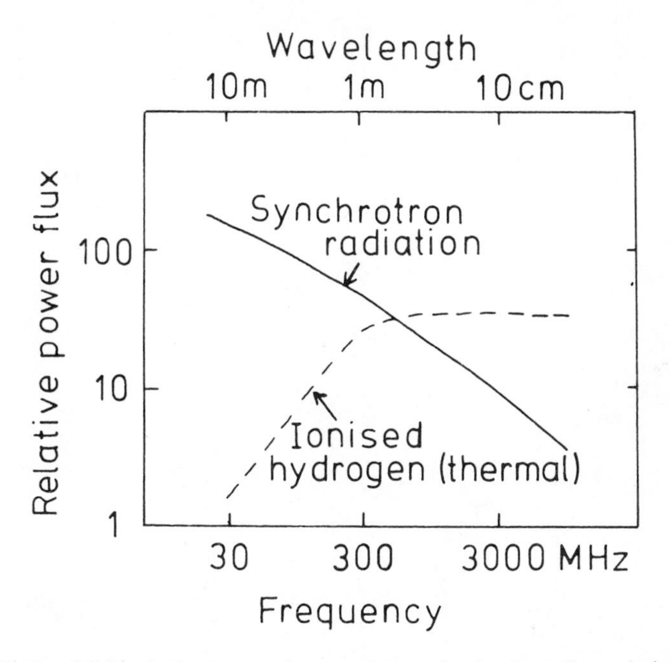

Figure 6.2 Typical spectra of components of galactic radio emission.

In the zone of ionised hydrogen, which is closely confined to the galactic disk, Reber's first postulated explanation of the origin of galactic radio noise stands correct. In a notable survey, one of the first to be undertaken with the Dwingeloo 25 m radio telescope after its completion, Westerhout (1958) at $\lambda = 21 \cdot 6$ cm recorded many individual emission regions near the galactic plane.

At metre wavelengths, the non-thermal radiation predominates, and many low-resolution maps have been drawn. Typical is the map in Figure 6.3 (b) at $\lambda = 1 \cdot 5$ m compiled by Dröge and Priester (1956) with a 17° beamwidth in which their results for the northern hemisphere have been combined with previous Australian results for the southern hemisphere. Figure 6.3 (c) shows a more detailed map of part of the Galaxy ob-

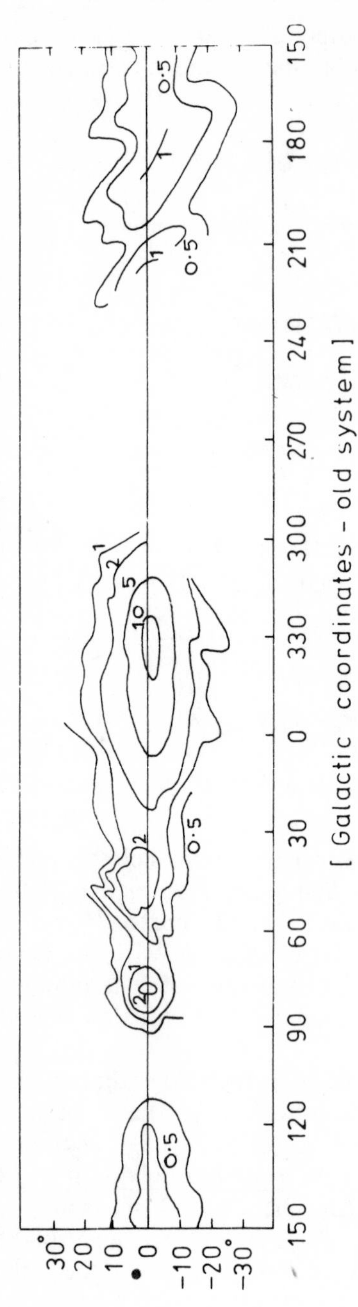

Figure 6.3 Maps of galactic radio emission.
(a) First map at $\lambda = 1{\cdot}9$ m (*After Reber, 1944*)

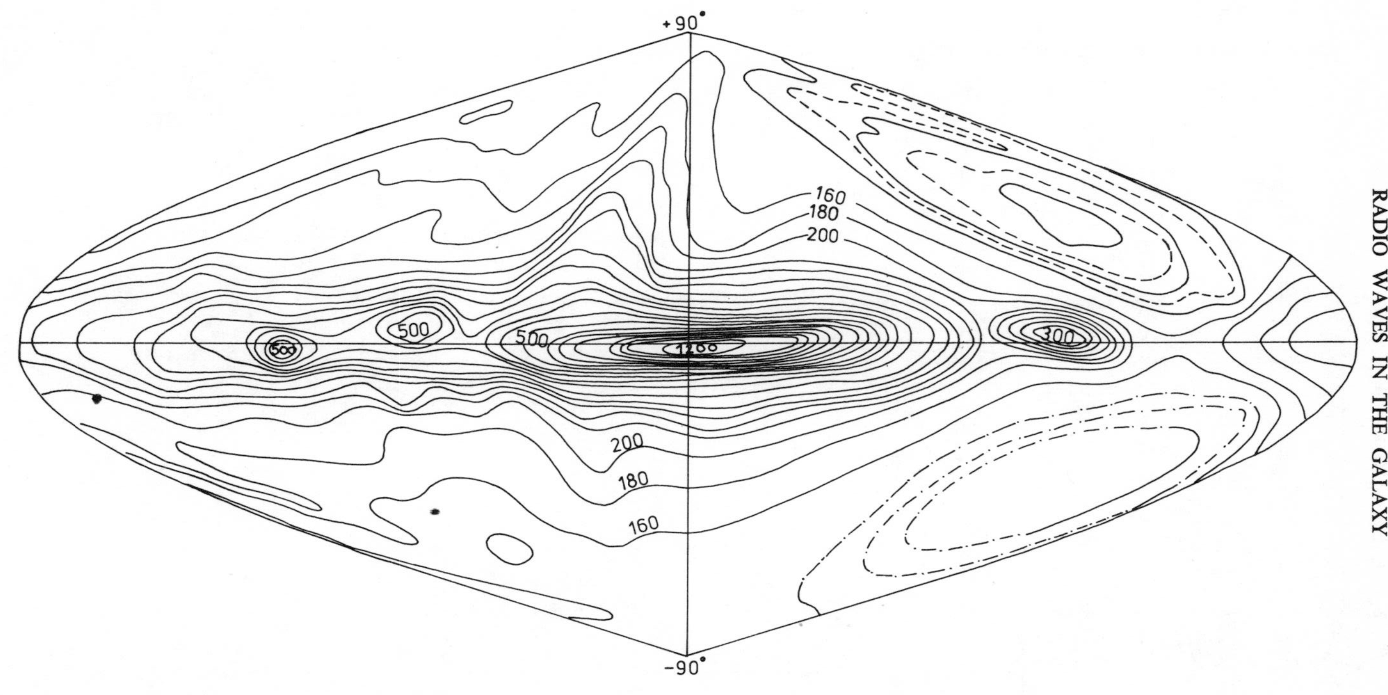

(b) Map at $\lambda = 1 \cdot 5$ m (*After Dröge and Priester, 1956*)

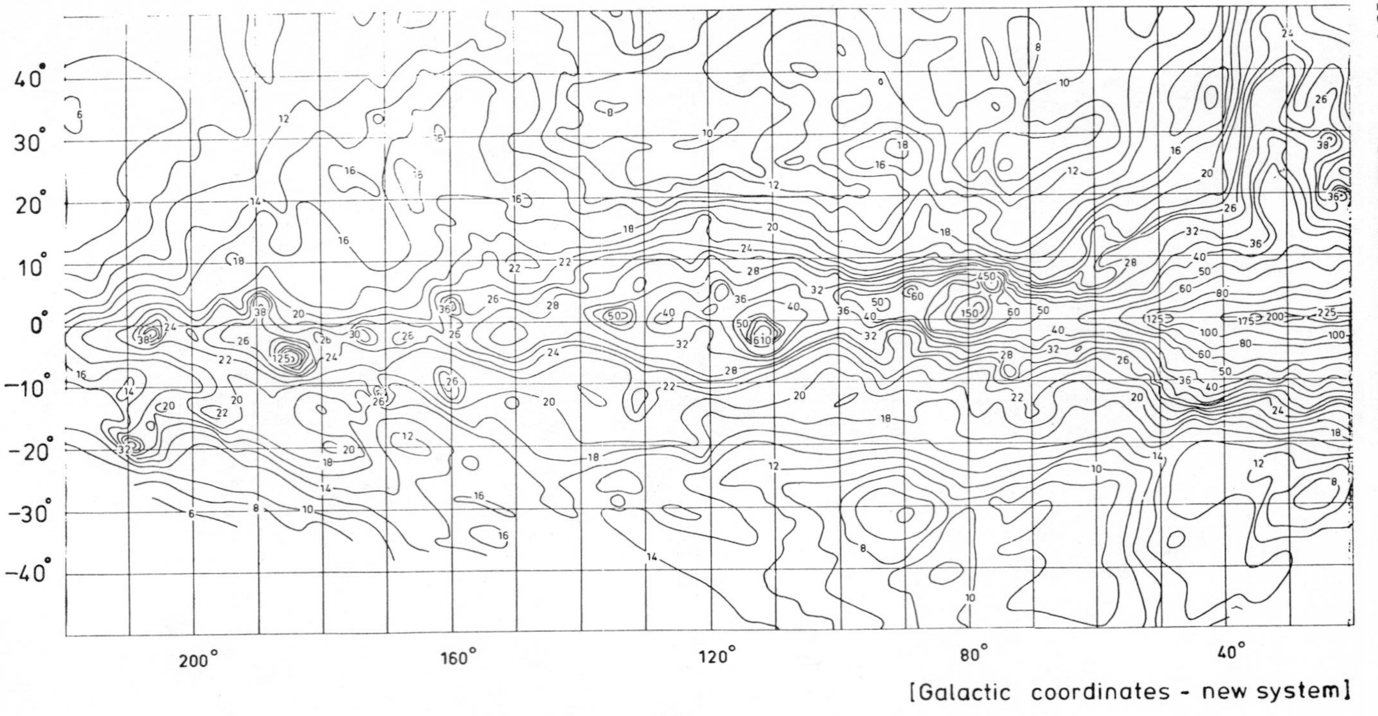

(c) Map at $\lambda = 75$ cm (*After Seeger et al, 1965*)

tained with the Dwingeloo radio telescope with a 2° beam-width.

The non-thermal emission, although concentrated towards the galactic plane, also extends to high galactic latitudes. After making appropriate allowance for the contribution of discrete sources, Shklovsky in 1952 propounded cogent arguments in favour of the existence of a galactic corona or halo, which may be pictured as an immense spheroidal distribution enveloping the Galaxy. The idea was reinforced by the wide extent of radiation which had been observed around Andromeda. The galactic halo appeared to be well-confirmed by studies in 1955 by Baldwin at Cambridge, and by Mills in Australia who found that scans at $\lambda = 3{\cdot}5$ m taken across the galactic equator revealed a strong narrow ridge of about 5° width, sometimes called the non-thermal disk, superimposed on a widespread distribution attributed to the halo. In 1959 Mills examined the non-thermal ridge in direction along the galactic equator, and noted a succession of steps which could be associated with spiral arm structure.

Subsequent attempts to be more specific about the properties, or even existence, of the galactic halo have met with some frustration. Surveys with high resolution have revealed more detail, particularly in the form of loops and spurs, but the problem is to know how much is localised in the immediate neighbourhood, and how much is large-scale galactic structure. One prominent spur, the North Polar spur, was evident in the earliest surveys, and is clearly visible in the map shown in Figure 6.3 (b). Another feature, the Cetus arc, was delineated by Large *et al.* (1962) in a survey made with the Jodrell Bank 250 ft telescope. More loops and arcs have become apparent, and it has been suggested that they might be shells of old supernovae. Consequently, it has seemed possible that the whole of the apparent galactic halo could be accounted for in this manner.

The success of the synchrotron theory in explaining the intensity and spectrum of non-thermal radiation naturally induced radio astronomers to look for characteristic linear polarisation. Such attempts were thwarted for many years, and it was assumed that the initial polarisation must be lost during transmission through complex field patterns. Eventually, in 1962, Westerhout *et al*, observing at 408 MHz ($\lambda = 73$ cm) with the 25 m Dwingeloo radio telescope, successfully traced out a region of polarised emission from the Galaxy. Confirmation was soon forthcoming from Wielebinski and Shakeshaft at Cambridge. In the southern hemisphere, Mathewson and Milne (1964), combining Dutch observations with their own obtained with the 210 ft radio telescope, recognised that the polarised regions lay predominantly along a wide band in a great circle, which they interpreted as polarised radiation from the local spiral arm.

The concentration of radio brightness towards the galactic centre almost certainly represents an agglomeration of sources, both thermal and non-thermal. A particularly strong non-thermal source Sagittarius A, appears to coincide with the centre. I will now proceed to discuss the various types of discrete radio sources that can be found within the Galaxy.

SUPERNOVAE

The recognition of powerful radio emission from the Crab Nebula added one more remarkable property to these exciting supernova remnants. As described in Chapter 2, Bolton, Stanley and Slee in 1949 first suggested that the radio source Taurus A coincided with the Crab Nebula.

Subsequent measurements of position and angular size in 1951–2 by Smith at Cambridge and Mills in Australia clinched the agreement more precisely. The identification provided a pointer to the type of astrophysical phenomenon one might look for in identifying intense radio sources, namely, some unusual explosive activity.

In 1953 Shklovsky suggested that in the Crab Nebula not only the radio emission but also the optical radiation might arise from the synchrotron process. Such an idea was supported by the continuity of the spectrum from radio to optical wavelengths. Further confirmation came from the detection by Dombrovsky (1954) of linear polarisation in the optical radiation. Although no radio polarisation had been observed, it was realised that complex magnetic fields and differential Faraday rotation could depolarise the received radio waves. Short wavelengths are least affected by Faraday rotation, and in 1957 when reliable techniques for determining small percentages of polarisation had been developed, Mayer, McCullough and Sloanaker, observing with the 50 ft NRL radio telescope, detected 7% linear polarisation at 3 cm wavelength. This notable achievement, the first observation of linear polarisation in a discrete radio source, left no further doubts about the validity of the synchrotron hypothesis of radio emission.

Following the recognition of the Crab Nebula as a radio source it was natural to seek other identifications with supernova remnants. The supernova of 1572 recorded by Tycho Brahe was bright enough to be visible in broad daylight at that

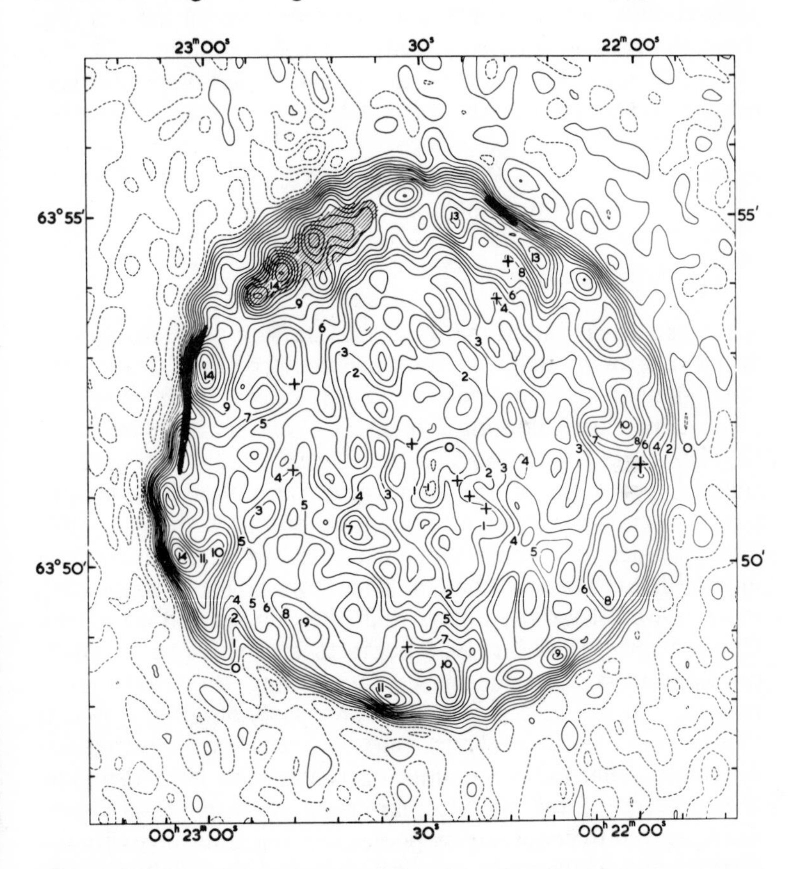

Figure 6.4 Radio map of the remnants of Tycho Brahe's supernova at $\lambda = 21 \cdot 3$ cm (*After Baldwin, 1967, IAU Symp. No.* 31, p. 352)

time, but all that now remains are a few faint wisps. In 1952, Hanbury Brown and Hazard detected a radio source in the direction of the supernova with the 218 ft fixed dish at Jodrell Bank. The identification of Cassiopeia A, described on p. 54, provided further proof that supernova remnants are powerful sources of radio emission. An interesting sideline on Cassiopeia A was Shklovsky's study in 1960 of the expected rate of decline of the field and particle energy. His theoretical calculations predicted that the radio source should weaken by 1 or 2% per annum. Careful measurements of flux density were then made by Högbom and Shakeshaft (1961) at Cambridge, and comparison with measurements in earlier years indeed showed that the radio flux was declining by about 1% per annum.

A supernova shell, as it expands against the interstellar gas, roughly takes a spherical form, which has been remarkably portrayed in the detailed maps obtainable by aperture synthesis. A striking example is the radio map in Figure 6.4 of the remnants of Tycho Brahe's supernova.

FLARE STARS

It was soon apparent that if normal stars emitted radio waves no more strongly than the Sun they would be too far away for radio detection. However, Luyten's discovery in 1948 of sudden increases of optical brightness of the red dwarf star UVCeti suggested the possibility of intense radio outbursts on stars. Other stars of the same type were found to exhibit irregular short-lived increases in optical output, sometimes up to 5 magnitudes. In comparison, solar flares are rarely discernible in total light. The idea that the flare stars might be immensely more powerful radio emitters than the Sun seemed attractive. Consequently, Lovell in 1958 began to organise a radio watch on flare stars, particularly UVCeti, using the 250 ft radio telescope at $\lambda = 2{\cdot}5$ m. A simultaneous photographic watch was essential to avoid chance bursts of interference being mistaken for stellar radio emission. Fortunately, in 1960, Whipple, the director of the Smithsonian Astrophysical Observatory, USA, agreed to cooperate by photographing flare stars with the Smithsonian network of cameras deployed for satellite tracking. The summation of radio recordings at the time of stellar flares, as described by Lovell, Whipple and Solomon (1963) provided unmistakable evidence of enhanced radio emission from flare stars. A similar radio watch started by Slee in Australia in 1960 with the aid of the Parkes 210 ft radio telescope at $\lambda = 73$ cm successfully demonstrated a clear coincidence of a single flare event with a radio outburst (Slee, Solomon, and Patson, 1963). Much subsequent work on flare stars has been carried out by Lovell in liaison with Russian astronomers. Flare stars, with their sudden violent release of energy, certainly present a fascinating astrophysical phenomenon. It appears, however, that although the flare stars may generate bursts of radio emission which can be a million times more powerful than those associated with solar flares, the total flux from all flare stars nevertheless amounts to no more than a small fraction, perhaps about ·1%, of the total radiation from the Galaxy.

PULSARS

No event in radio astronomy seemed more astonishing and more nearly approaching science fiction than the discovery of the clocklike radio pulses emitted from the sources now known as pulsars. The pulsating radio sources are best detected on a very large aerial at metre wavelengths with a receiver designed to record rapidly varying signals. It was a lucky chance that these instrumental requirements happened to be fulfilled during a Cambridge study of the rapid twinkling of radio sources of small angular size. A massive array of 2048 dipoles at $\lambda =$ 3·7 m covering more than four acres had been assembled for the investigation. When the survey of twinkling radio sources was being pursued in late 1967 by Miss Jocelyn Bell, she noticed some curious recurrent signals on the recording charts when the aerial pointed at a certain direction in the sky. After discussing the signals with A. Hewish, the leader of the project, they decided to install high-speed recorders to examine the waveform. It then transpired that the mysterious source emitted sharp pulses at such precise intervals, just over a second, that it seemed almost like a beacon designed to broadcast time signals. The fixed celestial direction of the source ruled out any chance that it might be interference from a terrestrial transmitter. The signals appeared so incredible that the Cambridge team pursued investigations for several weeks before releasing news of their discovery. Could the signals, they speculated, originate from a transmitter controlled by intelligent beings located elsewhere in the Galaxy? Such a possibility was discounted on several grounds. The signals varied irregularly in amplitude with no sign of a sensible code. Also, it could be assumed that if there were other civilisations they would exist on planets orbiting around stars, but the signals showed no periodic Doppler shifts attributable to planetary motion.

The first discovered pulsar was labelled CP1919 meaning Cambridge pulsar at right ascension 19 hrs 19 min. One property yielded a clue to the distance of the source; the pulses arrived slightly later at longer wavelengths. The dispersion in arrival times could be attributed to the scattering of radiation by interstellar electrons, the longer wavelengths being retarded the most. The delay in transit times gave an indication of the number of electrons in the path and hence an estimate of CP1919 at 400 light years away, a distance comparable with the bright stars. The next question was to decide the nature of the source since there was no obvious visual identification. If the pulses resulted from oscillation or rotation, their rapidity suggested a small, extremely condensed object. As it happened, previous theoretical work on white dwarfs, and on hypothetical neutron stars, pointed to a solution. In neutron stars, the protons and electrons of atomic nuclei would have coalesced into neutrons so crammed together that the star diameter would only be about 10 km. It is most noteworthy that the first paper by Hewish and his colleagues in February 1968 covered the basic facts and interpretation in such a comprehensive manner. A second paper several weeks later announced the detection of three more pulsars.

Radio astronomers elsewhere now joined in the search for more pulsars. Jodrell Bank quickly added more data on pulse shapes, dispersion in the galactic medium, and spectra. Lyne and Smith examined the polarisation and were the first to find a very high percentage of linear polarisation, in some cases up to 100%. A spate of papers appeared in scientific journals, principally from the groups at Cambridge and Manchester in Britain, at NRAO and Arecibo in USA, and at CSIRO and

Sydney in Australia. The maximum energy in the pulse spectrum lies around 100 to 400 MHz, and the 1000 ft dish at Arecibo and the new Mills Cross at Molonglo proved particularly well suited for the pulsar search. By 1970 over 50 pulsars had been listed, their periodicities mostly in the range $\frac{1}{4}$ to 2 seconds, with pulse widths typically about 5% of the pulse period.

The various theories proposed tended to settle around the hypothesis expounded by Gold (1968) that a pulsar is a fast-spinning neutron star with a very high magnetic field of the order of 10^{12} gauss resulting from the contraction of the original field of the collapsing star, and that the relativistic speeds and conditions at the surface produce a "lighthouse" beamed emission. One prediction following this hypothesis, that pulse repetition rates might be expected to be gradually slowing down, was admirably confirmed by subsequent observations.

Further support came from the recognition of an association with supernova remnants since it was believed that exploding stars could degenerate into neutron stars. In late 1968 Large, Vaughan and Mills at Sydney found a southern hemisphere pulsar near the centre of an extended radio source Vela X believed to be the debris of a supernova explosion. At that time it was the most rapid pulsar known, with a period of 89 milliseconds. Soon afterwards, Staelin and Reifenstein, searching with the 300 ft transit telescope at NRAO, found a pulsar in the Crab Nebula, NP0532, for which Comella *et al* measured the period, only 33 ms, making it the fastest and therefore probably the youngest pulsar known. Next came the exciting announcement in 1969 by Cocke, Disney and Taylor at the Steward Observatory, Arizona, that the Crab pulsar was simultaneously emitting light pulses, and that it coincided with the star which Baade had previously identified as the exploded supernova. What Baade had assumed to be continuous light emission from a star was in fact an optical pulsar. Then other workers using rocket launched X-ray equipment detected coincident X-ray pulses. Curiously enough, the X-ray power was about 100 times that of the optical pulsar, which in turn was about 100 times the radio power. The Crab Nebula, a comparatively near and recent supernova in a clear field of view has provided a unique opportunity to study a most important and fascinating astrophysical phenomenon.

RADIO STARS

So far I have discussed two types of radio source associated with exceptional stars, namely flare stars and neutron stars (pulsars). In recent years four other kinds of radio star have been recognised. The successful radio detection of several classes of stars has been summarised in an article by Hjellming and Wade[1] who played a substantial part in making the observations at NRAO where the three-element interferometer proved well-suited for the measurements which demanded very accurate directions coupled with high sensitivity.

I have previously indicated that any stars detected at radio wavelengths must radiate far more strongly than the Sun. Of course, certain stars are much larger than the Sun, and greater size obviously helps. Red supergiants having diameters several hundred times greater than the Sun are examples, and the detection of spasmodic radio emission from Betelgeuse (α Orionis) has been claimed. Radio observations of the red supergiant Antares (α Scorpii) proved especially curious and interesting, for it is part of a double system and the position measured with the NRAO interferometer revealed that the radio source in fact coincided with the blue dwarf companion, Antares B, of the red giant Antares A. Comparatively young novae have been found to comprise yet another class of radio source. Such stellar explosions, although less dramatic than supernovae, are characterised by a sudden increase of visual magnitude and the expulsion of an expanding shell of hot ionised gas, producing thermal radio emission which has been detected in certain instances of recent novae, such as Nova Delphini (1967) and Nova Serpentis (1970). Finally, a number of X-ray stars, like Sco X–1 and Cygnus X–1 have been found to be sources of variable radio emission. The disclosure of such a variety of different types of radio stars opens a new and promising field of research, made possible through the development of radio telescope systems of high sensitivity and accuracy. The term "radio star", abandoned in the 1950s when discrete

[1] Hjellming, R. M., and C. M. Wade, *Science*, **173,** 1087 (1971).

sources were identified with nebulae, now at last comes into its own correct usage.

EMISSION NEBULAE

In contrast with spectacular radio sources like supernovae and pulsars, let us now return to examine the classical radiation from hot ionised gas, mainly hydrogen (H II), surrounding the very hot O and B stars. Many such regions in the Galaxy have been recognised visually as emission nebulae, the best known being the Great Orion Nebula. As indicated by the spectrum of thermal emission from an ionised gas, (*see* Figure 6.2), the best chance of detecting such sources and distinguishing them from the non-thermal background is by observing at short wavelengths. With the 50 ft NRL radio telescope at $\lambda = 9{\cdot}4$ cm, Haddock, Mayer and Sloanaker (1954) undertook a series of observations of discrete sources, and in addition to recording sources like Cygnus A, Cassiopeia A and Taurus A, they succeeded in detecting bright emission nebulae such as the Great Nebula in Orion and the Omega Nebula in Sagittarius.

Soon further emission nebulae were recognised as radio sources, and the range of observation extended both to shorter and longer wavelengths, so that it became possible to derive the classically expected spectrum and to deduce electron temperatures and densities from radio data. In 1962, with the NRAO 85 ft radio telescope, Menon succeeded in mapping the radio distribution of the Orion Nebula and the Rosette Nebula at centimetric wavelengths. When large aerial systems like the new Mills Cross came into operation, the mapping of emission nebulae was also achieved at longer wavelengths. As the ionised hydrogen is mainly confined to the galactic disk, and is commonly in close juxtaposition with dark obscuring dust clouds, there are many H II regions for which adequate information is obtainable only by radio.

RADIO SPECTRAL LINES

The detection of new radio lines, coupled with the attainment of high angular discrimination, marked a new and exciting phase in the study of the gaseous condensations where stars are born. The close association of hydrogen and other gaseous constituents, dust clouds, and hot O and B stars, is of course no mere chance since it is here that pockets of matter contract under gravitational forces to become sufficiently hot and compressed to form new stars. The radio observations of molecular lines, and of lines of excited atoms, have brought forth some very surprising results, and an unexpected wealth of detailed data, promising to yield highly significant information relating to the activity and evolution of the medium in regions of star formation.

When spectral lines of the OH molecule at about 18 cm wavelength were first detected in 1963, nothing extraordinary was then apparent in the observations which were reasonably in accord with theoretical expectations. The observations represented a notable step in the radio detection of molecular lines, but perhaps the most interesting aspect of this initial investigation was the introduction to radio astronomy of a new method of spectral analysis. Weinreb, who was then a research student at the Massachusetts Institute of Technology, perfected the radio application of the method, autocorrelation spectrometry, based on well known mathematical principles of statistical analysis. By digitising the signals that constituted the received radio noise and computing the autocorrelation function, and then taking the Fourier transform, the power spectrum was derived. The practical accomplishment of the method was achieved through the available advances in fast sampling and computing techniques. (For instance, statistical analysis of a received signal covering a bandwidth of 1 MHz requires a sampling rate of 2 million times per second.) Weinreb, in cooperation with colleagues at MIT, successfully applied the method to detect two absorption lines at about 18 cm wavelength from interstellar OH when looking towards the radio source Cassiopeia A with an 84 ft radio telescope at Lincoln Lab., MIT. The Australian group led by Bolton, using the more conventional type of spectral analysis with narrow channel receivers, soon followed by observing OH in absorption looking towards Sagittarius A with the 210 ft radio telescope. The observation of OH lines was not unexpected, for although the abundance of molecular gases in the galactic medium was known to be low, the eventual detection of weak radio lines had been anticipated by the theoretical work of Shklovsky (1953), Townes (1957), and others.

It was not long before puzzling properties appeared. At CSIRO, Robinson *et al.* (1964) found anomalous intensities, and Doppler shifts unrelated to the motion of neutral hydrogen. But the most amazing results emerged in 1965 from the detection of OH in emission from the vicinity of H II regions, reported by Weaver's team at Berkeley, California, showing intense, narrow lines. The complexity of the line profiles led them to postulate the existence of an unidentified line which they labelled "mysterium". Subsequent observations showed that all the lines were OH, but in highly perturbed states. The strong intensities and narrow bandwidths of the emission lines undoubtedly indicated some kind of maser action, but even now, in 1972, the precise processes and pumping mechanisms are open to speculation. OH emission has been detected not only in H II regions but also near infrared stars, and in certain non-thermal galactic sources, and it seems probable that different activating mechanisms may apply, perhaps collisions in very dense regions, infrared radiation in some, ultraviolet in others. Amongst other exceptional properties, Weinreb *et al.* (1965), observing with the 120 ft Haystack radio telescope, noticed linear polarisation in some parts of the H II source, W3. Subsequently Davies *et al.* (1966) at Jodrell Bank, and Barrett and Rogers (1966) at NRAO found circular polarisation a more common characteristic, some components of OH sources being up to 100 per cent circularly polarised.

Clearly, the next task was to call on the resources of interferometry to find the positions and sizes of the OH sources. For instance, with the Jodrell Bank–Malvern interferometer it was demonstrated that some were less than $0''\cdot 1$, corresponding to linear sizes less than 10^{-3} parsec and brightness temperatures $> 10^{11}$ K. (Davies *et al.*, 1967). When transcontinental baselines were brought into action, it was shown that OH sources, like that in W3 for example, comprised a number of separate components, some with angular sizes $< 0''\cdot 005$ and linear sizes $<$10AU. Such concentrated OH regions, presumably associated also with aggregates of other molecules and dust particles could well be protostars, preliminary stages of star formation.

The detection of OH lines has been followed by a spate of discoveries of radio molecular lines. The success can be attributed to the perfection of methods of radio spectral analysis, and of radio telescopes and receivers in the microwave bands where molecular lines are more readily distinguished, and to the fortuitous circumstances of appropriate maser actions to boost the radio intensities. In comparison, the optical detection of interstellar molecular lines has lagged behind, for no interstellar molecules had been detected optically since CH and CN lines were observed in the 1930s.

In 1968, a team from Berkeley, University of California, detected ammonia (NH_3) emission lines near $\lambda = 1 \cdot 2$ cm looking towards the galactic centre with a new 20 ft radio telescope, and in 1969 they observed water vapour emission lines at $\lambda = 1 \cdot 35$ cm from several sources. Another group using the NRAO 140 ft radio telescope detected interstellar formaldehyde, H_2CO. In 1970, the new 36 ft NRAO telescope came into action, and CO, CN and HCN were detected at mm wavelengths. Then a heavier organic molecule, HC_3N was found at cm wavelengths in observations with the NRAO 140 ft telescope, and later in 1970, methyl alcohol CH_3OH. The list mounts year by year. The existence of polyatomic molecules was contrary to expectations for it was not generally believed that complex molecules could survive in the prevailing environment. It is remarkable to find in space the existence of molecules known to be important in biochemical reactions. The study of molecular abundances, the chemistry of the reactions producing organic molecules, and the evolution of gas clouds in space, have received a great impetus from the radio line discoveries. A tremendous field of radio astronomical study has been opened, combining astrophysics and astrochemistry, focused on a key fundamental phenomenon, the formation of stars.

During the rapid advance of molecular line research considerable progress was also made in the studies of H II regions, both by high resolution investigations revealing compact regions and superbright "knots", and also by observations of atomic excitation lines. Kardashev pointed out in 1959 that the radio lines of excited hydrogen at high quantum numbers should be observable in H II regions. Transitions of the type $n \rightarrow n - 1$

would have the highest probability, and the first detected line, corresponding to the transition $n = 105 \rightarrow 104$, was observed in the Orion and Omega nebulae by Dravskikh and Kolbason in 1964. Helium lines were detected by Lilley *et al*, in 1966, and a fair number of atomic transition lines, mostly of hydrogen and helium, have since been reported by various workers. The lines provide further data on the parameters of H II regions, and a valuable guide to their kinematics.

The task that lies ahead is to obtain more information of the detailed structure and properties of galactic nebulae to elucidate the processes of evolution. The spatial, radiative, and collisional relations have to be sought between dust clouds, condensations of neutral hydrogen, concentrations of OH and other molecules, protostars and new stars, and the zones of ionised and exciting gas. The problems are complex, but fascinating and challenging. The prevalence of obscuring matter limits the optical probing of these intriguing nebulae. It is radio astronomy with its penetrability, its wealth and variety of data, that offers most scope for future advance and breakthrough in our knowledge.

CHAPTER 7

Radio Galaxies, Quasars, and Cosmology

The study of external galaxies has been one of the principal objectives of radio astronomical research. It has a great bearing on our understanding of the evolution of galaxies and cosmology; and the application of the synchrotron theory has proved an invaluable guide to the interpretation of the energy of the sources.

The subject is too vast to trace fully the sequence of progress in two decades. In this short account, I can only indicate briefly the range of aspects encompassed by this field of research. After the first identifications of certain radio sources with galaxies, it was evident that much painstaking work lay ahead. The obvious requirements were for accurate positions to aid optical identifications, radio intensities and spectra to

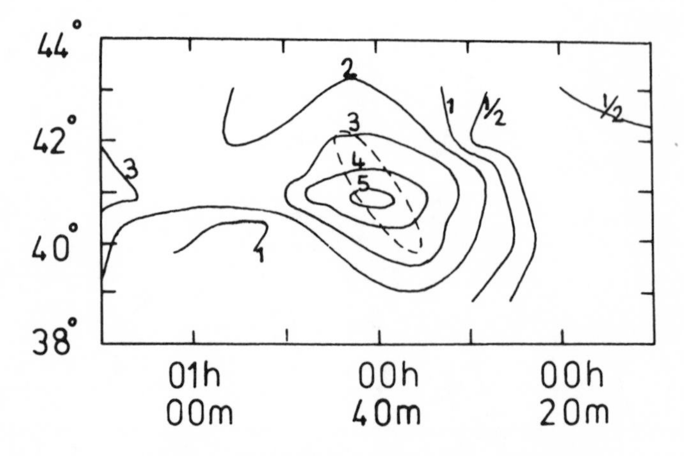

Figure 7.1 Radio maps of Andromeda Nebula.
(a) First map at $\lambda = 1{\cdot}9$ m (*After Hanbury Brown and Hazard, 1951*)

reveal emission processes, polarisation to indicate magnetic fields, and radio structures to compare with visible galaxies.

NORMAL GALAXIES

Normal galaxies were naturally among the first to be examined for their radio characteristics. They are the galaxies which fit the standard optical patterns of Hubble's classification of spirals, ellipticals and irregulars, excluding those with exceptional features in size, structure or brightness. Early series of observations of normal galaxies in the northern hemisphere were made at Jodrell Bank by Hanbury Brown and Hazard (1959, 61). Figure 7.1 contrasts their simple 1951 map of

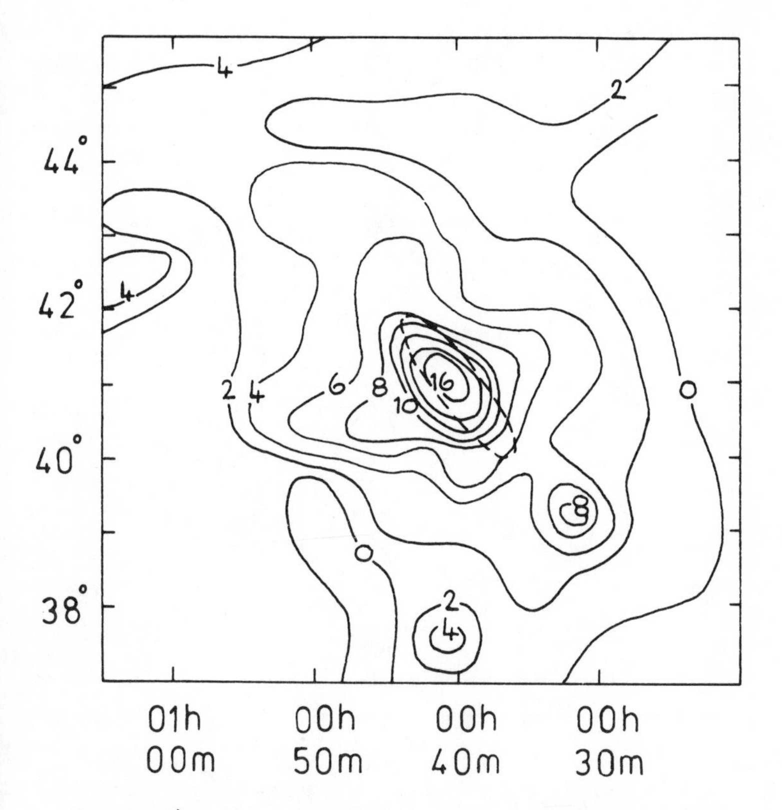

(b) Map at $\lambda = 73$ cm (*After Large, Mathewson and Haslam, 1959*)

Andromeda, with the more detailed and complex distributions subsequently obtained by others with finer resolution. In the southern hemisphere, Mills (1955) observed several normal

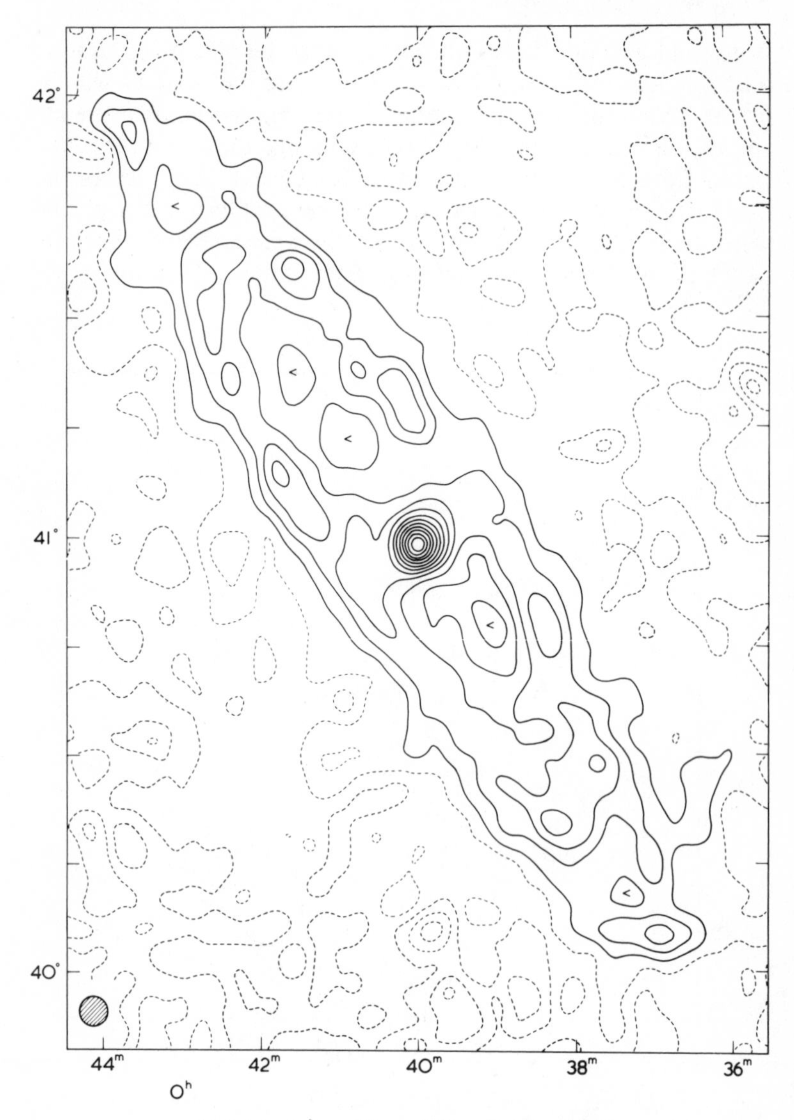

(c) Map at $\lambda = 73$ cm (*After Pooley, 1969*)

galaxies with his Cross aerial giving a 1° beamwidth at $\lambda = 3 \cdot 5$ m enabling him to plot radio contours of the Magellanic Clouds. Normal galaxies have continued to receive close attention, for instance at NRAO by Heeschen and Wade (1964), Heeschen (1970). The 21 cm hydrogen line emission from the Magellanic Clouds was plotted by Kerr, Hindman and Robinson of CSIRO in 1954, and the neutral hydrogen content and distribution has since been examined in many normal galaxies.

I will not attempt here to pursue a discussion of the many valuable results, but will merely state that for the most part it has been found possible to associate typical radio emission with the different classes of galaxy. The general inference has been that normal galaxies do not exhibit any excessive radiation compared with spiral galaxies like our own, so in this sense their radio emission also is classified as "normal". For these objects, radio methods have been able to add to the optical fresh information to invigorate the research on the classical types which constitute the major population of galaxies in the universe.

RADIO GALAXIES

Most interest has been focused on the galaxies with far greater output, like Cygnus A, that are known as radio galaxies. In this field, radio astronomy has led the way in disclosing many extraordinary galaxies and astrophysical phenomena of vast energy and activity that must surely possess great evolutionary significance. Despite the great knowledge emanating from radio data, it must be noted that in several respects interpretation leans heavily on optical identifications, since optical redshifts provide the only means of estimating great distances. In addition, the comparison of the radio structures with the optical is a key to understanding the astrophysical processes involved. The outcome of research on radio galaxies can be illustrated most graphically by a few selected examples. But first I will summarise some of the essential groundwork on the general properties of radio galaxies, conclusions that have emerged through years of patient study.

RADIO SPECTRA

With continually improving accuracy of flux measurements, the spectra of large numbers of sources have been reliably graphed over the radio band. Early estimates of intensity were inaccurate by as much as 50%; the elimation of erroneous values was achieved gradually through the perfection of noise

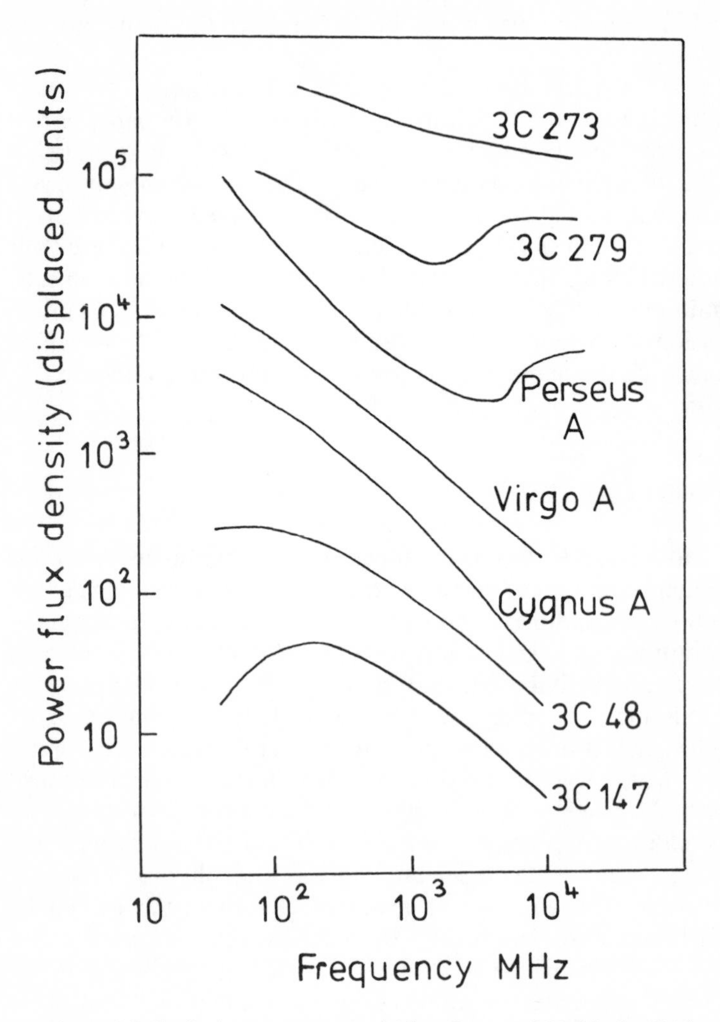

Figure 7.2 Typical source spectra (*After Dent and Haddock, 1966*)

standards, aerial calibration, and repeated observations with different apparatus. Once values had been established with sufficient certainty for a few sources, they became celestial standards for comparative measurements of other sources. The same approach has been applied to the determination of directions, where a few well testified locations have been used as reference indicators of positions in the sky. The comparative method avoids the necessity of continually calibrating directly the absolute parameters of observing equipment.

In 1963, Conway of Jodrell Bank, Kellermann of NRAO, and Long of Cambridge, felt that the spectral characteristics of many sources were sufficiently well established to warrant a collaborative attempt to publish definitive data. Although their list of 160 spectra stands as a landmark in the tabulation of radio source spectra, subsequent measurements have led to amendments and additions. Particularly important has been the extension of spectra to centimetric wavelengths where previously there was a paucity of information. For example, Dent and Haddock (1966) at Michigan, extended many spectra to 8000 MHz ($\lambda = 3{\cdot}75$ cm), and a great deal of data down to $\lambda =$ 11 cm has been accumulated in surveys by the Parkes 210 ft radio telescope. Typical source spectra are illustrated in Figure 7.2, and I shall discuss later the marked curvature apparent in certain spectra. In principle it was clear that the synchrotron theory satisfactorily accounted for the source intensities and spectra.

POLARISATION

Considering that linear polarisation is a characteristic of synchrotron emission it seemed strange that so many years had to elapse before polarisation was detected in a radio galaxy. Of course, it was appreciated that Faraday rotation and complex magnetic field patterns could drastically reduce the received overall polarisation, especially at longer wavelengths. Once more it was the NRL group that scored the first success when in 1962 Mayer, McCullough and Sloanaker using the 50 ft radio telescope at 3 cm wavelength observed 8% linear polarisation from Cygnus A. The polarisation at 10 cm wavelength was found to be no more than 0·5% so accounting for previous

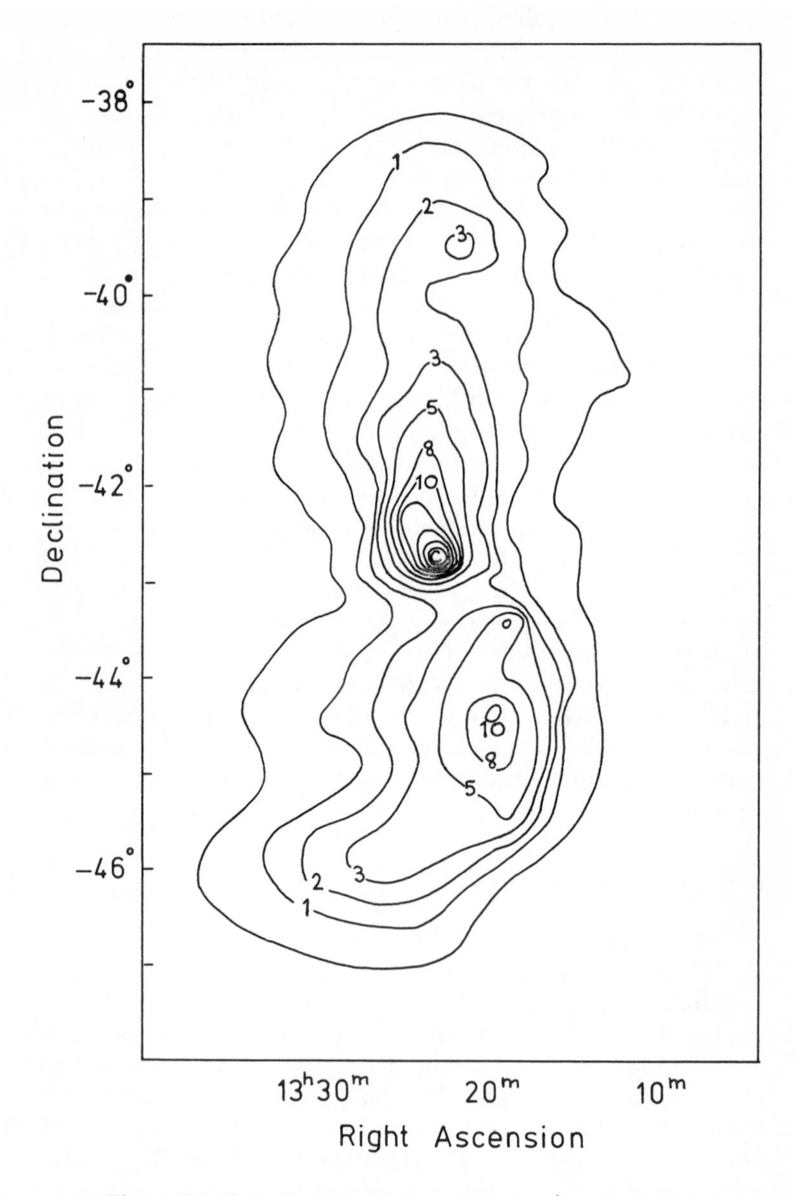

Figure 7.3 (a) Radio map of Centaurus A at $\lambda = 21 \cdot 3$ cm

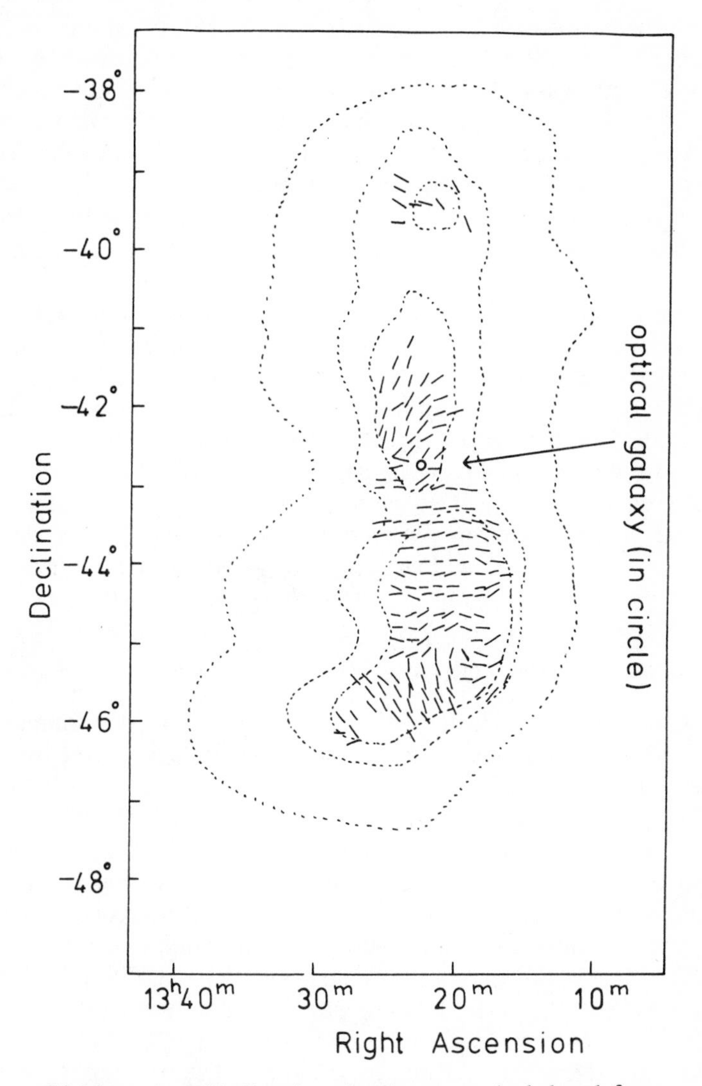

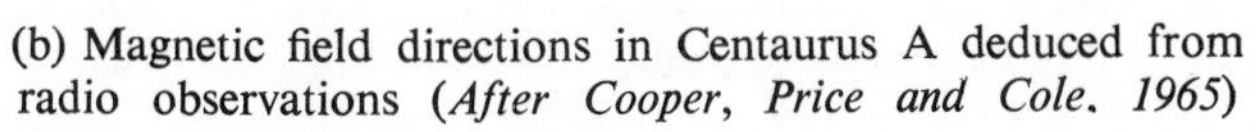
(b) Magnetic field directions in Centaurus A deduced from radio observations (*After Cooper, Price and Cole. 1965*)

failures to detect polarised radiation at longer wavelengths from this source. Close on the heels of NRL, the detection by Bracewell, Cooper and Cousins at CSIRO of about 15% linear polarisation in the central part of Centaurus A at 10 cm wavelength was one of the first accomplishments of the Parkes 210 ft radio telescope. In 1963, Gardner and Whiteoak at CSIRO observed the polarisation of many sources and showed how measurements at different wavelengths enabled the direction of the magnetic field at the source to be deduced by extrapolation. They also concluded that much of the Faraday rotation experienced in observing extragalactic sources is imposed on the radio waves as they pass through our Galaxy. The radio galaxy, Centaurus A, with its large angular size, lent itself to detailed examination, and Cooper, Price and Cole (1965) were able to map with the 210 ft radio telescope the structure of the magnetic field as shown in Figure 7.3. For distant sources, of small angular size, the early measurements referred to the overall polarisation, indicating a predominant trend in field direction. Later studies have gleaned more data about the magnetic field structures of radio galaxies, clearly a matter of great import in understanding their formation.

SOURCE STRUCTURES

Generalised conclusions about the nature and development of radio galaxies must be based on the properties of large numbers of sources and the manner in which they can be classified into characteristic groups. A valuable step towards this goal was accomplished when Maltby and Moffet (1962) applied the Cal. Tech. variable spacing interferometer to the study of source structures. They measured the interference fringe patterns in both amplitude and phase at baseline spacings up to 1560λ. It was fortunate that by this time the Jodrell Bank team had sufficiently perfected the long baseline technique to examine fringe visibilities up to 60 000 λ spacing, and were thus able to extend the analysis to the more distant galaxies subtending smaller angular diameters (Allen, Hanbury Brown and Palmer, 1962). These two complementary surveys of the morphology of radio sources were substantially in agreement, showing the most common type to be the double source with well separated components of large dimensions compared with the visible parent galaxy. Some sources were found to be multiple, others core-and-halo structures. Actual sizes spread over a wide range, the large components being mostly between 10 000 and 100 000 light years with several times greater separation. The very large double or complex regions extending far beyond the confines of the optical galaxy strongly implied that high energy particles and fields producing the radio emission have been thrown outward and expanded from a past explosion in the optical galaxy. Compact components may also be evident, especially at the shorter wavelengths, usually in the vicinity of the galaxy. The fact that many sources, or principal components, were still too small in angular diameter to be resolved, prompted further extensions in baseline to a million wavelengths (the Jodrell Bank–Malvern interferometer), and then to many millions of wavelengths in transcontinental interferometers. Measurements of sources of small angular dimensions will be discussed later. Meanwhile, with the aid of aperture synthesis methods such as those developed by Ryle and his team at Cambridge, many sources of intermediate angular size were being mapped in some detail.

As the number of identifications increased to comprise a representative sample it became possible to distinguish the types of optical galaxies that form strong radio sources. It was found that radio galaxies do not fall into Hubble's grouping of normal galaxies. Based on Morgan's classification of more unusual galaxies, an analysis by Matthews, Morgan and Schmidt (1964) showed that the powerful radio sources were particularly associated with exceptional optical objects, usually exhibiting strong optical emission lines, some showing remarkable jets or other unusual features.

CYGNUS A

The high intensity of power flux from the first discovered radio source, Cygnus A, made it a special target for the investigation of its radio structure. Owing to its great distance of about 550 million light years, its angular subtension was only of the order of a minute of arc, and the attainment of such resolution in the early 1950s severely taxed the ingenuity of radio

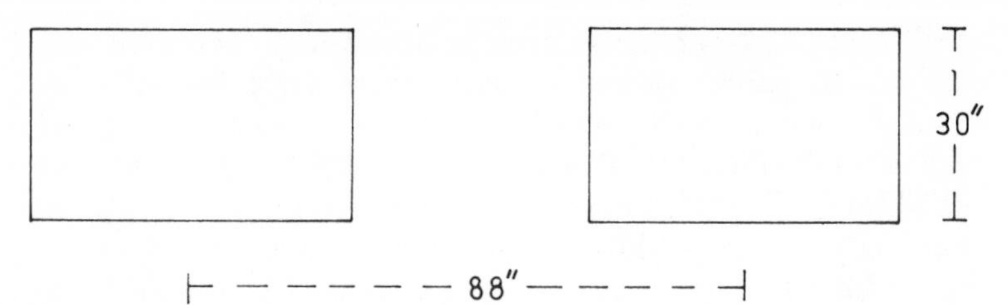

Figure 7.4 (a) First radio structure of Cygnus A
(*After Jennison and Das Gupta, 1953*)

astronomers. In 1953, Jennison and Das Gupta at Jodrell Bank, measuring fringe amplitudes at spacings up to 4 km, succeeded in deriving the basic double structure shown in Figure 7.4 (a). Of course, later refinement of interferometry and aperture synthesis provided a far more detailed picture as illustrated in Figure 7.4 (b) showing the contour map obtained in 1969 by Mitton and Ryle at Cambridge.

For a few years after Baade and Minkowski's identification of the associated optical galaxy it was firmly believed that its double shape and presence of highly excited emission lines indicated two galaxies in collision. Although the hypothesis of colliding galaxies seemed to fit a few subsequent identifications, it was eventually realised that neither the available energy nor the probability of collisions was adequate to explain the numbers and radiated power of radio galaxies.

A fundamental problem in the interpretation of radio galaxies has been to account for their tremendous energy, which for the most powerful sources demanded a catastrophic transformation of up to 10^{61} ergs of galactic energy into the energy of fast particles and magnetic fields required to generate the observed radio emission. Theoretical papers by G. R. Burbidge from 1956 onwards focused attention on the problem presented by the transference of so much energy to the particles and fields, which could attain as much as $0{\cdot}1\%$ of the total mass energy ($E = mc^2$) of the whole galaxy. Amongst the various proposals, Shklovsky (1960) suggested very frequent supernova explosions in the galactic nucleus, and Burbidge (1961) invoked the idea of a chain reaction of supernova outbursts. Hoyle and Fowler (1963) put forward the hypothesis that following the gravitational collapse of a galaxy the ensuing explosion could

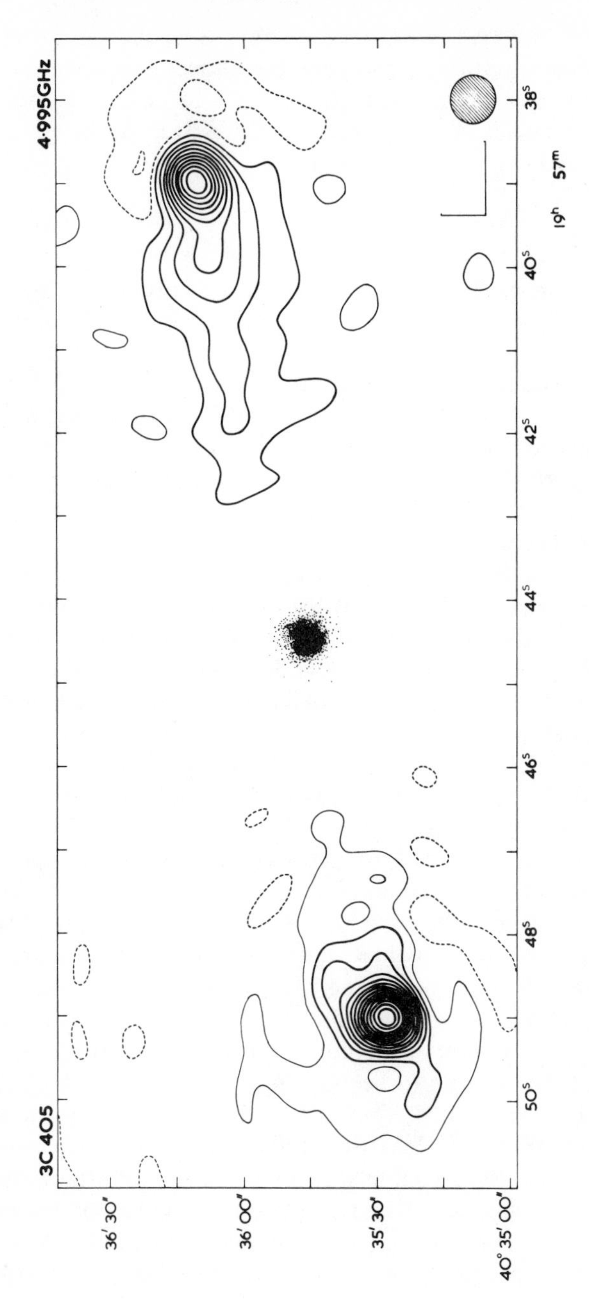
4·995GHz
3C 405
38s
19h 57m
40s
42s
44s
46s
48s
50s
36' 30"
36' 00"
35' 30"
40° 35' 00"

account for the vast release of energy. Speculation on this fundamentally important problem continues. Optical studies have increasingly emphasised the evidence of violent events in the galactic nuclei, as described by Burbidge, Burbidge and Sandage (1963).

QUASARS

Exciting and baffling as the radio galaxies proved to be, the discovery of quasars in 1963, arising from combined observations by radio and optical astronomers utilising to the full the available resources of instrumental techniques, revealed the most astonishing of all extragalactic objects. The quasar story really began in 1960 when the position of the radio source 3C48 was determined with an accuracy of about 5″ by Matthews by means of the Cal. Tech. radio telescope interferometer. Measurements of angular size by Palmer's team at Jodrell Bank had already shown the source to have a very small diameter, less than 1″. Guided by the accurate position, Sandage photographed the region in a 90 minute exposure with the 200 inch Palomar telescope, and observed that the direction coincided with an unusual type of blue star. It seemed that a true "radio star" had at last been found. It was certainly an unusual star, for Greenstein and Sandage were unable to recognise the spectral lines. In the next two years a few more radio sources, including 3C 273, were similarly identified with peculiar blue stars. In 1962, a dramatic turn of events opened a new phase in the realisation of the nature of these mysterious objects. In that year the path of the Moon happened to cross the radio source 3C 273 three times, on April 15, August 5, and October 26, thus providing the opportunity for Hazard, Mackey and Shimmins, observing with the 210 ft CSIRO radio telescope, to make a very accurate determination of the position of the source. Hazard had been a foremost protagonist of the lunar occultation method which allowed source positions to be deduced to within 1″. Hazard and his colleagues (1963) found 3C 273 to comprise a double source, and one part which they designated as component B had a very compact core which seemed to coincide with a 13th magnitude blue star. With the 200 inch Palomar telescope, Schmidt (1963) confirmed the identification and at the same time noticed a faint jet apparently associated with the other radio component (A). The startling revelation came with the recognition of the spectral lines as emission lines with a redshift of 0·158 indicating a very distant and extraordinary extragalactic object of stellar appearance. From its optical magnitude, and its distance of 1500 million light years (according to Hubble's law of redshifts), it could be inferred that the intrinsic optical luminosity must be over 100 times greater than that of any normal galaxy. Greenstein and Matthews immediately decided to re-examine the spectrum of the object that had been identified with 3C 48, and they at once realised that the spectral lines corresponded to an even larger redshift of 0·367. Other similar identifications quickly followed. So the three years' puzzle of the curious "radio stars" was solved by this exciting discovery of these hitherto unknown, distant, tremendously luminous objects of stellar appearance, the quasistellar radio sources, or quasars as they were soon called, emitting radio powers comparable with the strong radio galaxies. To add to their extraordinary characteristics, Smith and Hoffleit (1963) found that the "star" corresponding to 3C 273 had been visible on Harvard photographic plates taken over a period of 80 years, and had fluctuated in brightness. Matthews and Sandage also noted optical variations in 3C 48, with large changes in brightness in less than a year. Such rapid variations in magnitude implied objects no more than a light year in diameter!

The search for quasars, and the study of their properties in comparison with radio galaxies, became the most intriguing problem in extragalactic research. Many quasars were found during the following years. It was evident that not only did the optical radiation originate from a very small region, but that the radio source, although larger, also contained concentrated components. The endeavour to determine the sizes and structures of compact sources at great distances presented an acute test of the skill of radio astronomers. Great enterprise was shown in the attempts to deduce radio source sizes, including some extremely interesting indirect methods. The most straightforward approach has been by interferometry and lunar occultations. The less direct but nevertheless valuable additional methods have been by logical deductions that could be drawn from variability, scintillations and spectra. Let us now consider the respective merits of the different methods.

LUNAR OCCULTATIONS

The utilisation of the occultation of radio sources by the Moon to derive source positions and structures was considered as early as 1950 by Getmanzev and Ginzburg who discussed the diffraction pattern expected when a source was covered by or emerged from the lunar disk. During the next ten years, radio astronomers availed themselves of the opportunities afforded by occultations only on few occasions. For two conditions had to be fulfilled to permit frequent application of the method. A large collecting area was required to observe a weak source and record the Fresnel diffraction pattern, and the source must be tracked continuously throughout the occultation. A large steerable radio telescope is obviously the principal instrumental requirement, and the advent of the 250 ft Jodrell Bank telescope made an occultation programme a far more feasible proposition. So it was not until 1960 that Hazard deliberately began to exploit the technique first with the Jodrell Bank 250 ft radio telescope (starting with the occultation of the extragalactic source 3C 212), and then with the 210 ft Australian telescope where the method enjoyed its most notable triumph in the observations in 1962 of the occultation of 3C 273 which led directly to the discovery of quasars. Scheuer, of Cambridge, was at this time working at CSIRO, where he wrote a well known paper on the analysis and restoration of occultation curves. He demonstrated that resolution is not limited to the angular size of the first Fresnel zone ($\sim 10''$); in fact, accuracy better than $1''$ is readily attained. The method soon became more widely adopted and can easily be applied over a wide range of wavelengths. H.M. Nautical Almanac Office at Herstmonceux has most helpfully supplied predictions and astronomical data. Up to 1970 over 50 radio sources had been examined, most of the results being obtained with the 250 ft Jodrell Bank telescope, the 210 ft Australian telescope, the 1000 ft Arecibo telescope, and the 140 ft telescope at NRAO. The chief disadvantage of the method is that one has to await the occurrence of appropriate occultations, and even so, they are limited to the equatorial band traversed by the Moon.

INTERFEROMETER OBSERVATIONS

The most general and successful method of observing small sources has been by interferometry. The gradual, continual improvement in directional precision and resolution achieved over the years by interferometric methods has been, in various adaptations including aperture synthesis, the most vital technical advance in radio astronomy. Most impressive has been the factor of improvement in measurements of angular dimensions in 25 years—from fractions of a degree to hundredths of a second of arc and less—for in the early days of radio astronomy no one dreamed that such resolution would ever be attainable.

The accurate determination of source positions requires very stable interference fringes and fortunately does not demand baselines on the geographical scale necessary for the analysis of small structures. In fact, up to 1970, most positional measurements have been achieved with separations of up to about 1 km allowing the aerials to be connected by cables. By the early 1960s, variable spacing interferometers employing two large parabolic reflectors were in operation at the Cal. Tech. observatory at Owens Valley, and at the Royal Radar Establishment, Malvern. In my group at RRE we began by using a spacing of 750 m at 610 MHz giving a lobewidth of 140″. The phase of the interference pattern could be held steady within 1 %, so that directions of sources could be ascertained with an accuracy of about 1″ in both Right Ascension and Declination (Adgie and Gent, 1964, 1966). Radio astronomers were now beginning to compete in accuracy on equal terms with optical astronomers.

In subsequent years, determination of radio positions with similar accuracy in both coordinates were made elsewhere, for instance at NRAO by Wade (1970). Of course, identifications can be attempted, often quite successfully, on the basis of radio directions of lower accuracy than 1″. The precision of the Parkes 210 ft radio telescope was increased with the aid of a dual feed system to compare signals in two overlapping beam directions, and Merkelijn (1969), Bolton and Wall (1970), derived radio positions at 2700 MHz with an estimated accuracy of about 10″ to 15″, leading to many identifications with radio galaxies and quasars. Throughout all the work of identification, the Palomar Sky Survey plates taken with the 48 in Schmidt telescope have been the invaluable guide to optical positions. Without doubt, radio positional accuracy of 1″ represents a desirable goal for establishing identifications, and such accuracy becomes a necessity when the 200 inch Palomar telescope and long exposures are brought into action to discern very faint optical objects. The identification of radio sources is the basic link of the whole subject with optical astronomy.

Quasars, as well as other radio sources having components of extremely small angular diameter, stimulated the pursuit of long base systems with continually increasing baselines. By 1962, the Jodrell Bank team led by Palmer had extended the baseline to 60 000λ. Anderson, Palmer and Rowson were able to demonstrate that 3C 295, the most distant identified radio galaxy, with a redshift of 0·47, consisted of two components separated by about 4″. The next stage was the cooperative venture between the groups at Jodrell Bank and RRE, Malvern, with the MK2 radio telescope at Jodrell connected by a triple-hop radio link to a 25 m parabolic reflector at Malvern. Even with baselines at 11 cm and 6 m of 1 and 2 million wavelengths to measure angular diameters down to 0″·025, some quasars such as 3C 273B, as well as the core of the radio galaxy 3C 84 (Perseus A), remained unresolved (Palmer, Gent *et al.*, 1967).

The next step in extending baselines became a race between groups in Canada, USA, Australia, and England to devise systems dispensing with the radio link, employing instead extremely stable local oscillators controlled by atomic clocks at the two sites, and recording output data on magnetic tapes which could later be correlated by playing the tape recordings together into a computer. I will not attempt to allocate credit for the idea of such a method, for it must have long been in the minds of many who considered future interferometer systems when several techniques, atomic clocks and fast recording and computing, had advanced to the pitch where they could be incorporated in a practical scheme. In the event, the Canadians won the race over USA by a short head by first bringing the system into successful operation in 1967. In Canada, Broten *et al*, observing at 448 MHz ($\lambda \approx 67$ cm) with a baseline of 3074 km, equivalent to 4·6 million wavelengths, obtained strong fringes on the two quasars 3C 273 and 3C 345. Soon afterwards Bare *et al.* in USA also brought into operation an interferometer on the same principle.

By 1970, several experimental systems had been set up with various baselines traversing the Earth, the longest baseline of all to date being 10 592 km established by NASA Space Tracking radio telescopes between Goldstone, California and Canberra, Australia, representing 81 million wavelengths at $\lambda = 13$ cm, and interference fringes on many sources were clear evidence of components less than 0″·001 in angular size, as Kellermann *et al.* (1970) have described.

A subsequent series of observations with a variety of spacings epitomises both the extraordinary resolution attained by radio methods, and provides an exemplary demonstration of international scientific cooperation. As my account of the development of radio astronomy nears completion, the results of this investigation have just been published in the *Astrophysical Journal* **169**, 1 (1971) in a paper entitled "High resolution of compact radio sources at 6 and 18 cm". At 6 cm wavelength the longest spacing, 10 536 km, corresponded to 176 million wavelengths thus enabling the structure of compact components of many quasars and galaxies to be examined with an angular resolution down to 0″·0004. Even with this resolving power, several sources such as 3C 273 and 3C 279 contained powerful components of smaller angular size.

The collaboration is aptly illustrated by the list of authors and the research institutes they represent as follows:

U.S.A.:	K. I. Kellermann (NRAO and Cal. Tech.) M. H. Cohen and B. B. Shaffer (Cal. Tech.), B. G. Clark and J. Broderick (NRAO)
Sweden:	B. Rönnäng and O. E. H. Rydbeck (Onsala Space Observatory)
U.S.S.R.:	L. Matveyenko (Institute for Space Research) I. Moiseyev (Crimean Astrophysical Observatory) V. V. Vitkevitch (Lebedev Physical Institute)
Australia:	B. F. C. Cooper and R. Batchelor (CSIRO)

Such investigations are notable for the remarkable experimental results and for the collaboration between scientific groups without which success would never have been achieved. While competition and rivalry have played their part in the progress of research, most of the great advances in radio astronomy owe their success to cooperation between individuals and groups. Indeed, the scale of radio astronomy apparatus makes it inevitable that often many people will be involved in a particular experiment, and a stream of authors' names have consequently appeared on published papers. Sometimes the first mentioned name may be that of the leader, or more often the one who is alphabetically lucky, whilst "*et al*" frequently has to suffice in subsequent discussions to describe the long list of other contributors.

SCINTILLATIONS

Whenever scintillation is detected in radio astronomy it seems that a gem of knowledge is uncovered. In 1949 the apparent fluctuations of Cygnus A led to our knowledge of discrete radio sources, as well as to valuable data about irregularities in the ionosphere. Fifteen years later, in 1964, rapid scintillations noticed in a number of sources disclosed a wealth of information about the diameters of sources of very small angular size and on the properties of the solar wind which I discussed in Chapter 5. During a series of transit observations at $\lambda \approx 1 \cdot 7$ m at Cambridge, Clarke noticed that a few sources, such as the quasar 3C 147, exhibited continual scintillations. Systematic investigations of the scintillating sources undertaken by Hewish, Scott and Wills established that the fluctuations occurred when observing sources of extremely small angular size, less than about 1″, and were caused by irregular ionisation in the solar wind. The analysis by Hewish and his co-workers demonstrated that the study of amplitude scintillations provided a potent means of estimating the sizes of sources subtending only a fraction of a second of arc. The method has the merit of requiring only a large radio telescope to look at the sources. Cohen in USA became engaged in scintillation investigations from 1965 onward, utilising the Arecibo 1000 ft dish at the longer wavelengths, and the NRAO 140 ft telescope at short wavelengths. Very small limits could be placed on sources showing strong scintillation. For example, Cohen and Gundermann (1967) observing at 11 cm and 21 cm wavelength, were able to set an upper limit of 0″·005, and a corresponding linear size of 13 pc, to a strong component of the quasar 3C 279. Both the British and US workers demonstrated that model fitting could be successfully applied to infer source structures.

CURVED SPECTRA

Just as the study of source scintillations has been amazingly lucrative in estimating source sizes, it was equally unexpected to find in 1963 how much could be inferred about the size and properties of small intense sources simply from peculiar features in their radio spectra. The point is that if powerful sources have small dimensions, their surface radio brightness temperatures must be extremely high. In such circumstances, as explained by Le Roux (1961), self-absorption of the synchrotron emission can occur, causing a steep fall in radio output at longer wavelengths. The realisation that certain sources, such as the quasar 3C 147, exhibited this type of spectral turnover, and that in consequence the angular size could be estimated, was first pointed out in 1963 by the Russian theoretician Slish and independently by Williams at Cambridge. Again, the method has been particularly valuable in indicating compact sources subtending only a fraction of a second of arc. The calculation of source sizes depended on assumed magnetic fields; alternatively, as Williams (1966) demonstrated, an upper limit to the magnetic field could be extracted if angular sizes were already known from long base interferometer measurements. In this way he concluded that in the compact sources, equipartition between the magnetic energy and that of the relativistic particles often does not hold, the particle energy being far the greater.

VARIABLE SOURCES

Finally, we consider how the intrinsic variability shown by some sources imposes a limit to their possible size. The argument is so well known that it may safely be accredited to "Anon". If a source shows a marked alteration of emitted radiation in say one year, then since disturbances causing such changes cannot be transmitted faster than the velocity of light, it may be assumed that the source cannot exceed a light-year in diameter. The first clear evidence of radio variability in an extragalactic source was presented by Dent of the University of Michigan in 1965 when he reported that measurements at $\lambda = 3 \cdot 75$ cm of flux density of the quasar 3C 273 showed a 40% increase between 1962 and 1965. As mentioned on p. 161, optical variations had previously been found in the source, and for this reason Dent had maintained a close watch on its radio stability. In addition he noted changes of radio intensity in the quasars 3C 279, and 3C 345, and optical inspection now revealed variations in optical magnitude also. In 1965, the observations of variability were extended to mm wavelengths. Epstein, observing with a 15 ft parabolic reflector at Aerospace Corporation, California, recorded considerable changes of the intensity of 3C 273 and 3C 279 within a period of six months. Low utilised the 200 inch Palomar telescope for his measurements at 1 mm wavelength where he also found large variations of received flux.

It is curious how, in the course of radio astronomy, certain phenomena have so long remained unnoticed, and yet when discovered are readily apparent and accepted, as if a mist had suddenly cleared to open up a new vista in knowledge and opportunities for research. Polarisation and variability are two examples. Until the early 1960s radio astronomers firmly held the view that the extragalactic sources were randomly polarised radio emitters of constant intensity. Indeed it was felt that the dimensions of galaxies were so vast that no short-term variations could be envisaged. Once linear polarisation and variability had been detected however, the situation radically altered. It was soon recognised that many sources manifested similar properties. The new findings admirably justified the persistent pursuit of observations at very short wavelengths at NRL and Michigan, for it is at cm and mm wavelengths that these characteristics are most apparent.

In 1967, Aller and Haddock at Michigan reported the occurrence of changes in the radio polarisation of certain quasars, as well as in their intensity. Variability was not confined to quasars; for in 1966, Dent observed changes in the radio output of the Seyfert galaxy NGC 1275 (radio source Perseus A, 3C 84), and variations have since been recorded in other radio galaxies such as 3C 120, as described by Pauliny-Toth and Kellermann (1968).

It had become abundantly clear that many quasars and some radio galaxies contain a very compact, intensely energetic core, often associated with explosive activity. The plentitude of methods for estimating the sizes, structures, and characteristics of radio sources was beginning to set the stage for the portrayal of a representative sample of types and the manner of their evolution. Promising attempts have been made by Kellermann (1966) to interpret the changes in spectra of variable sources in terms of expanding outbursts, as indicated by the spectral turnover due to self-absorption moving from short to longer wavelengths. In a further paper, Kellermann and Pauliny-Toth (1969) pointed out that owing to inverse Compton scattering (high energy particles transferring some of their energy in collisions with photons), synchrotron radiation would be limited to maximum brightness temperatures of about 10^{11} or 10^{12} K. Evidence that repeated outbursts may occur in a source has been forthcoming from the changes in emission from the core, from changes in shape of the central components recorded by very long baseline interferometers, and from the multiplicity sometimes found in large components. Large double, multiple, and core-and-halo structures occur in quasars as well as in radio galaxies. Models to explain the expansion of the vast clouds of relativistic particles and magnetic fields have been put forward by Shklovsky (1960) and others. The synchrotron theory has remained undisputed as the principal process of radio emission. But the problems of the production of relativistic particles and their replenishment by repeated activity have prompted a great deal of speculation. Here is the province of theoretical research which I will not attempt to review, for there are at least as many theories as there are theoretical astronomers. The evolutionary relationships between the different types of galaxies, radio galaxies, and quasars, and the mechanisms of release of enormous energy into the relativistic particles and fields are the greatest enigmas. So much fresh information is becoming available through combined optical and radio observations, supplemented by data obtained by infrared, X-ray and other newly introduced techniques, to make one feel confident that during the 1970 decade the scene will become much clearer and an integral picture of these distant celestial objects and phenomena, their nature and development, will emerge.

SOURCE SURVEYS

The initial source surveys by Ryle at Cambridge and Mills in Australia have already been mentioned in Chapter 4. It is clearly a basic necessity to observe, list, and classify as many radio sources as possible by measurements of their positions, intensities and angular dimensions. Such surveys provide a starting point, whether for general statistical analyses, or for detailed investigation and identification of individual sources. In establishing reliable lists, the cross-checking and comparison of surveys by different research teams using various aerial systems has proved invaluable. At first, the surveys were made at the longer wavelengths where the non-thermal sources are most intense. Gradually the advantages of additional surveys at shorter wavelengths were appreciated, especially when it was realised that confusion limited the number of sources to about one per 25 beamwidths. The narrower beamwidths at shorter wavelengths, coupled with the improvements in receiver sensitivity when masers and parametric amplifiers were introduced, permitted larger numbers of sources to be tabulated in the decimetric and centimetric wavebands. Multi-frequency surveys were essential in any case to establish the spectra of the sources.

By 1970 many valuable and comprehensive surveys had been undertaken. At Cambridge, the 3C survey of 471 sources at $\lambda = 1 \cdot 9$ m, after further cross-checking and amendment, was republished in 1961 as the 3C Revised catalogue of 328 sources, constituting a well-known standard reference list. The next, more comprehensive survey 4C, the first extensive series of observations with the aperture synthesis system at $\lambda = 1 \cdot 7$ m listed some 5000 sources down to 2 flux units. A much deeper survey, 5C, down to $0 \cdot 025$ flux units, uses the one-mile radio telescope at $\lambda = 73$ cm and is intended for detailed inspection of selected areas of sky. This survey is necessarily limited to areas of special interest, for it takes 2 months to cover 10 square degrees and a complete sky coverage would occupy 2000 years.

In the southern hemisphere, the Mills Cross MSH survey at CSIRO of over 2000 sources at $\lambda = 3 \cdot 5$ m (published in 1958, 1960, 1961) has been followed in recent years by measurements at shorter wavelengths, down to 11 cm, with the 210 ft radio telescope to compile the extensive Parkes catalogue of sources. Many other notable surveys have been started or completed at radio observatories in various countries, for instance, at Ohio State University and at Vermilion River, University of Illinois,

USA, at the University of Bologna, Italy, and elsewhere. A large proportion of the catalogued radio sources remain unidentified and it seems reasonable to infer that many are very distant objects. The conclusion that most extragalactic radio sources are strong radio emitters at great distances has been skilfully argued by Ryle (1958).

COSMOLOGICAL SIGNIFICANCE OF SOURCE COUNTS

In general, the diversity in the properties of radio sources tends to defeat attempts to derive cosmological inferences from statistical analyses. For instance, if the radio sources all had similar dimensions the relation between observed angular diameter and distance could have provided a particularly revealing indication of the space-time geometry of the universe. As it is, the linear sizes are known to range from thousands to millions of light years. One test only has led to cosmological conclusions, the counts of radio sources. Ryle found that counts of radio sources at different received intensities showed an unexpected excess of faint radio sources, thus implying an evolutionary universe. Since the time of travel of electromagnetic waves means that we observe distant sources at past epochs, it was inferred that radio sources were more numerous in the past and perhaps more powerful also.

The verification of the precise relation between numbers and intensities has been a lengthy procedure, with Ryle steering the way, and the Australian group cautiously and sometimes effectively applying the brakes. If the number density is uniform and the space-time geometry Euclidean, the number of sources N with power flux density greater than S should be inversely proportional to $S^{1\cdot5}$. The Cambridge 2C survey gave an index of 3 instead of 1·5, and although these results published in 1955 suffered from confusion, Ryle's attention had been drawn to the significance of source counts. After years of surveys with refined methods ensuring the elimination of errors more precise conclusions have been substantiated, yielding an index of 1·85 over a wide frequency range. The derivation of N–S distributions was reinforced by an ingenious mathematical method of analysis devised by Scheuer in 1957 to extend the sampling of recordings

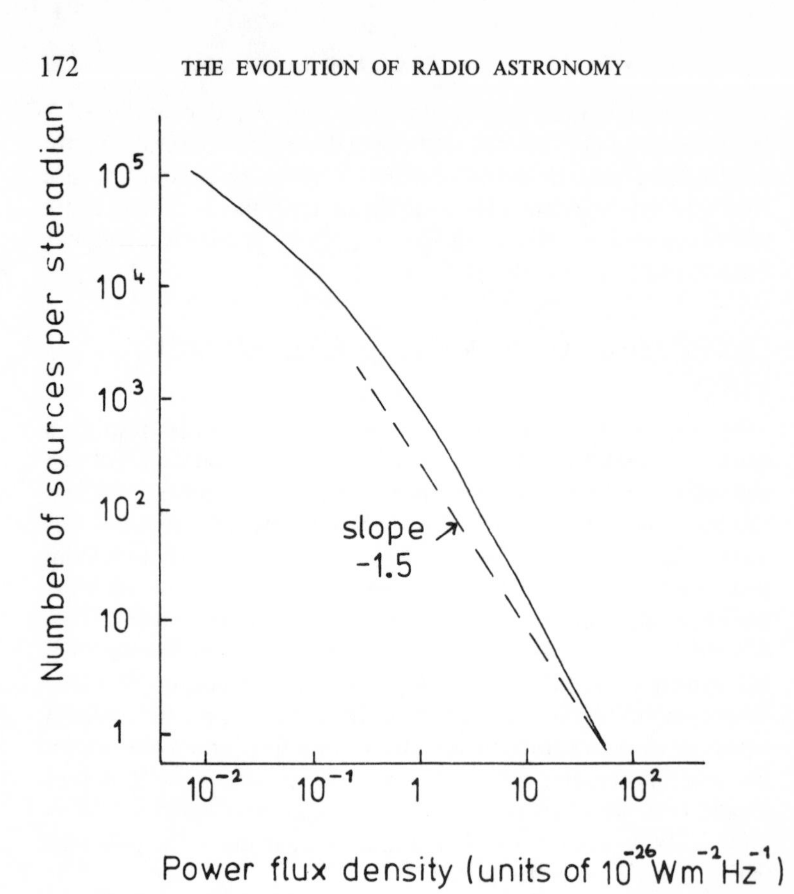

Figure 7.5 Source counts at $\lambda = 73$ cm (*After Pooley and Ryle, 1968*)

down to very weak intensities. Typical results of source counts at 408 MHz ($\lambda = 73$ cm) are shown in Figure 7.5. Plotting log N against log S, the magnitude of the slope is $1 \cdot 85$ down to $S = 4$ flux units. Thereafter the slope gradually diminishes at lower intensities. A decrease of slope is to be anticipated on any cosmology that is not static Euclidean. A sharp fall off in the apparent radio source population corresponding to the greatest distances, as discussed by Longair (1966), suggests a cutoff at redshifts of $z = 3$ or 4.

The Cambridge counts of radio sources at decimetre and metre wavelengths, and the evolutionary cosmological concepts that they imply, seemed to be well confirmed by the results of surveys elsewhere. Certain queries, however, have stemmed from the Australian centimetric surveys by Bolton and his colleagues with the Parkes 210 ft radio telescope suggesting lower values for the initial slope in the log N–log S graph at these short wavelengths. How the discrepancies are to be accounted for is not yet clear, but they have drawn attention to some interesting points. Sources with relatively flat spectra tend at the shorter wavelengths to become more prominent in comparison with other sources. Consequently, counts of sources down to a given flux level will comprise a rather different sample at shorter wavelengths. The question as to whether different log N–log S slopes can be attributed to particular types of sources, such as quasars and radio galaxies, has been raised previously by Véron (1966) and others. But it is not easy to separate into classes large numbers of sources when so many remain unidentified. Selection effects always bedevil attempts to ensure strict statistical sampling. Although Ryle has argued that in the complete log N–log S analysis it is unneccessary to differentiate between the various types of extragalactic source, one would have wished to have been able to sample different classes separately. It is generally accepted that Ryle's conclusions on the number-intensity relation are so thoroughly attested that their validity is unquestionable, but a few lingering doubts remain as to whether the last word on the subject has yet been said.

COSMIC BACKGROUND RADIATION

The events leading to the observation of microwave background radiation and the appreciation of its portent provide a fitting finale for my account of the progress of radio astronomy up to 1970. The interplay of participants seems almost to have been designed to portray the concluding act of an intriguing play. For this final scene, some of the early players are again on the stage. They are seeking, unknown to each other and with different motives, not a spectacular manifestation, but a minute clue that might hold the key to a crucial problem. For many years they have pursued their diverse roles. When they are brought together by one who previously figured in an exciting discovery, they collaborate to solve the elusive mystery. The solution turns out to be of tremendous significance; it holds a vital answer concerning the beginning of the universe. I will now proceed to relate the story.

The participants are research staff at the Bell Telephone Laboratories where the original discoveries in radio astronomy were made, Dicke and Peebles at Princeton University, and Burke who first found the intense radiation from Jupiter.

I described on p. 27, Dicke's construction of a sensitive microwave radiometer and his measurements of a finite sky background temperature. Although at that time he was unable to estimate the residual radio temperature very accurately, he then assumed that the background radiation might represent the contribution from all the galaxies in the universe. The paper by Dicke *et al.* appeared in 1946 in the *Physical Review*, **70,** p. 340 (1946). Dicke was unaware that 232 pages later in the same volume an article by Gamow considered the implications of the hypothesis that the universe began in an intensely hot compressed state, a primeval fireball at perhaps 10^{10} K from which it has expanded and evolved. The cosmological concept of a very hot beginning to the universe was not new, but Gamow was the first to elaborate the theory of the subsequent physical development and its influence on the formation of elements. His research students, Alpher and Herman (1949), estimated the residual temperature of the cosmic radiation after expansion at about 5 K. In the 1960s, Dicke and Peebles at Princeton, unaware of Gamow's work, were studying the formation of elements in an oscillating universe, which involved a hot, compressed "bounce", and again implied residual black-body radiation after the subsequent expansion. In contemplating whether the residual temperature could be measured they were at once reminded of Dicke's experiments of twenty years earlier to determine the sky temperature. Consequently, Roll and Wilkinson started to assemble at Princeton a more sensitive and stable type of Dicke radiometer connected to a horn aerial in order to measure the cosmic background temperature.

Meanwhile, only 30 miles away, Bell Telephone Laboratories, guided along an entirely different route by the demands of modern technology, had independently embarked on the identical measurement. When the Echo and Telstar projects were being engineered for radio transmissions via artificial satellites, the requirements for a low noise system had led Crawford at BTL to propose a 20 ft aperture horn reflector as the receiving aerial. During the Echo project, Ohm carefully evaluated at 11 cm wavelength all the sources of noise in the receiver system. A few degrees of unaccounted noise temperature were arbitrarily assigned to radiation from the horn. Later Penzias and Wilson, two radio astronomers who had joined the BTL staff, undertook to track down the unexplained noise. Since the excess noise showed no sidereal, solar, or directional variation, the horn aerial was suspected to be the source. But the 3 K excess temperature which they measured at 7·3 cm wavelength remained on re-erection of the horn after dismantling, cleaning, and elimination of possible imperfections. Having exonerated the horn aerial, and discarding after careful consideration all astronomical sources as a possible explanation, Penzias and Wilson were at a loss to know how next to proceed to solve the mystery of this minute but unmistakable additional noise. In a telephone conversation with Burke at MIT on another matter Penzias casually mentioned the enigma of the excess noise. Burke replied that he had just seen a preprint of a cosmological paper by Peebles of Princeton predicting that, if the universe evolved from a very hot compressed state, several degrees of cosmic background noise should be observable. Penzias immediately telephoned Dicke who responded at once by arranging a visit of the Princeton group to BTL. It now became clear that Penzias and Wilson at BTL had actually been measuring the cosmic background radiation that the Princeton team were preparing to look for with their own radiometer at 3·2 cm wavelength. Together they now published short papers in the *Astrophysical Journal* (1965). Six months later, the Princeton radiometer was completed and Roll and Wilkinson (1966) were able to confirm the observation of the 3 K background radiation. The next step was to extend the spectral range of the measurements to verify the expected thermal black-body spectrum. Support came from the Cambridge measurements of Howell and Shakeshaft (1966) at $\lambda = 21$ cm. Further indirect evidence was provided by optical observations of cyanogen (CN) absorption lines, since it could be calculated that a radiation intensity corresponding to about 3 K at $\lambda = 2{\cdot}6$ mm must be responsible for the excitation of the observed

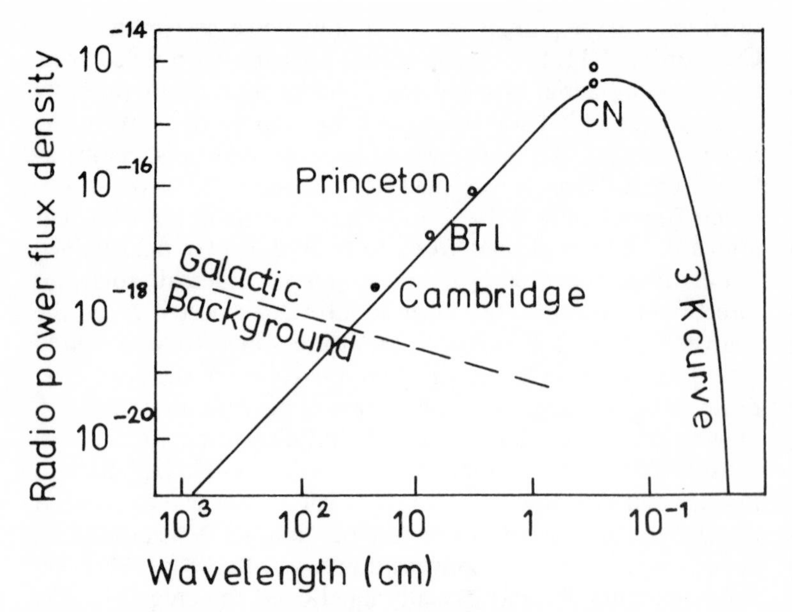

Figure 7.6 Initial measurements (by 1967) of the cosmic microwave background shown on the 3 K blackbody radiation curve.

lines. Microwave measurements of the radiation at other wavelengths soon provided excellent confirmation of the isotropic thermal background radiation corresponding to a temperature now more accurately assessed at 2·7 K. The spectrum is illustrated in Figure 7.6. It is a great tribute to the sensitivity and accuracy attained by radio methods that they have been instrumental in establishing so elegantly an observation of such fundamental importance concerning the history of the universe.

CHAPTER 8

The Scope of Radio Methods in Astronomy

Radio methods have so greatly advanced our knowledge of the universe that they are now an integral part of astronomical science. Vital new information has been acquired in almost every branch of astronomy and astrophysics. This has covered such diverse topics as cosmology, the evolution of galaxies, magnetic fields and high energy particles, the structure of the Galaxy, supernovae, emission nebulae, the solar atmosphere, sunspots and flares, the interplanetary medium, the Moon and planets, and so on. Yet the prospect of such tremendous progress by radio methods had been quite unforeseen. Forty years ago, in the early 1930s, the first astronomical observations of radio emissions evoked only a fleeting attention from astronomers. Even twenty five years ago, when a glimmering interest had been awakened by new discoveries, many regarded the subject as little more than an interesting sidetrack in the progress of astronomy. We must therefore pose the questions: why were the potentialities of radio astronomy not recognised earlier; and why did the extensive advance by radio methods so surprise both physicists and astronomers?

The long drawn-out interval between the experimental demonstration of radio waves by Hertz in the late 19th century and the rise of radio astronomy in the mid 20th century is, I think, attributable to a variety of factors. The first was the discouraging influence of the early failures to detect radio waves from the Sun. The attempts were handicapped by the insensitivity of radio detectors and ignorance of the ionospheric barrier to long wavelengths. The negative results disheartened radio physicists from further endeavours, expecially as it was suspected that radio emission from the Sun would be no more than the weak tail of thermal radiation from the 6000 K photosphere.

The reasons why so little regard was paid to Jansky's discovery in 1932 of radio waves from the Galaxy, and to subsequent observations of Jansky and Reber, are partially explicable as follows. Firstly, radio seemed to hold little promise of attaining high resolving power. The early plots of the radio distribution obtained with beamwidths greater than 10° were relatively crude representations of the Milky Way. Since radio waves are a million times longer than light waves it was assumed that radio could never compete with optical resolution. Secondly, the occurrence of radio waves from the Galaxy was at least partly accounted for by Reber's suggestion that the emission might originate from ionised interstellar gas in accordance with classical concepts of thermal radiation. Admittedly it had been noted that the intensity at decametric wavelengths was about ten times stronger than might be anticipated. Nevertheless there appeared neither sufficient precision in observation nor inexplicability in theory to suggest any far-reaching impact on astronomy.

At that time, moreover, most astronomers and physicists were unfamiliar with radio techniques. Advances in radio science were continually and rapidly being transformed into the technicalities of radio communication, and the gulf between the specialised "know-how" of the radio engineer and the interests and instruments of the astronomer could not easily be bridged. The factors which I have outlined above render the long hiatus in the development of radio astronomy more readily comprehensible.

The technical progress during the Second World War in electronics, radar, radio communication, and detection devices, brought a revolution in their applications to industry and science. Many physicists and research scientists had been involved in the development and deployment of radar, so that after the war they were admirably qualified to exploit the potentialities of the technical advances and to demonstrate how radio techniques could be adapted to radio astronomy. Great strides had been made in radio sensitivity leading to the exciting discoveries and achievements described in earlier chapters, pointers to new possibilities for research. It is true that the same progress would have been made eventually had there been no war. Civil aviation alone would have promoted the demand for radar. Broadcasting and television would similarly have spurred advances in radio and electronics. But the pressures of war greatly accelerated the rate of technical progress, and the transferences of scientists from one field of research activity to another undoubtedly had a beneficial outcome. Favourable conditions prevailed for launching the new ventures in radio astronomy.

We have yet to answer the second question concerning the amazing success of radio astronomy. Why have radio scientists found so many fruitful avenues of basic astrophysical research, as well as entering every few years an arena of serendipity revealing astonishing discoveries?

The explanation can be found, I think, partly in the ease of production of large numbers of radio photons and partly in the fine sensitivity of radio receivers. The quantum energy of a radio photon, $E = h\nu$ is small because radio occupies the low frequency portion of the electromagnetic spectrum. Phenomena producing energetic particles or photons can often either stimulate processes giving rise to low energy photons, or decay into low-energy forms of activity—an inherent trend dictated by the Second Law of Thermodynamics. The synchrotron mechanism is an outstanding example of high energy particles directly releasing radio emission. If we consider the other extreme, the radiation emitted by exceptionally cold bodies like the outer planets or the background radiation from space at 3° above absolute zero, only at infrared and radio wavelengths is any emission observable. We conclude that there is a strong probability that radio emission will accompany astrophysical processes whatever their initial energy scale; and that as radio photons possess small quantal energy they will occur in large numbers.

When we consider the detection of radio waves we find that the theoretical limit is set by the reception of radio photons rather than by their intrinsic energy. Radio receptors are therefore fundamentally extremely sensitive means of measuring low power. In practice, of course, the theoretical ideal is never achieved. Despite the cooling of receiving devices, such as the maser in liquid helium, it is impracticable to screen out every extraneous influx of noise. Nevertheless, receiver noise temperatures as low as 20 K can be attained in practical systems, and by long integration the limiting noise level can be reduced to less than a thousandth of this value. Consequently radio occupies the most sensitive part of the electromagnetic spectrum. Some years ago it was estimated that the total energy that had been picked up by all the radio telescopes in the world would only raise the temperature of a spoonful of water by about a millionth of a degree. Yet the sensitivity of radio receivers has enabled us to acquire a stupendous amount of astronomical information. We come to the conclusion that the strong likelihood of radio waves being generated by astrophysical processes and the high sensitivity of radio detectors, are factors which greatly enhance the radio observation of phenomena.

Why were the astronomical potentialities of radio not appreciated earlier? I do not think that before the 1940s anyone seriously contemplated the wide-ranging possibilities. Scientists are customarily preoccupied with their current research problems. Most scientific progress is not planned far ahead, but is the outcome of a gradual process of evolution. In consequence there is an opportune time for the development of new lines of research dependent on prevailing circumstances and pressures, which include not only theoretical ideas that may call for experimental test and verification, but also on the availability of techniques often prompted by industrial or military needs, as well as on the patronage of influential leaders of scientific policy and the subsequent organisation of financial support. Certainly, the progress of radio astronomy has been cultivated by the fulfilment of these conditions.

The knowledge gained by radio astronomy can be divided between those areas where the role of radio has been to supplement basic information derived by optical astronomy, and those aspects where previously unknown and predominantly invisible phenomena have been revealed by radio. We may illustrate the first category, for example, by the added knowledge of visible surfaces of the Moon and planets, or of the solar chromosphere, or of flare stars, where complementary data has been gleaned by radio methods. The second category comprises an impressive list, revealing a new range of astrophysical phenomena. Examples include the Jovian radiation belts, the interplanetary plasma, the galactic distribution of neutral hydrogen, zones of high energy particles and fields in radio galaxies, quasars, pulsars, maser emission of molecular lines, and so on. In some instances, as in quasars, the radio observations have acted as a guide and pointer to focus the attention of the optical astronomer onto new and exciting visible phenomena.

It is not only the variety of phenomena disclosed by radio that is so striking. Radio methods possess two especially advantageous qualities. One frequently mentioned is the ability of radio waves to "see" through the clouds and mist of the terrestrial atmosphere and through the interstellar dust and fog that causes so much visual obscuration. Freedom from interstellar obscuration is particularly valuable for instance, in deriving galactic structure by means of the 21 cm hydrogen line. The other merit of radio lies in the wide range of wavelengths that it covers. Optical wavelengths comprise one octave only in the electromagnetic spectrum (λ from 0·4 to 0·8 microns) while the radio band covers at least 10 octaves (λ from 0·1 cm to 10 m). The virtue of the wide range is especially evident in astronomical regions where absorption and emission may alter by orders of magnitude as the observing wavelength is changed —from millimetres to centimetres, decimetres, metres and decametres. An example is the solar atmosphere; emission at mm waves arises from levels near the photosphere, at cm waves from the chromosphere, at metre waves from the corona. Simply by changing the wavelength, radio can probe throughout the whole extent of the solar atmosphere. Or consider Jupiter, at mm waves the radiation is thermal emission from the surface, at decimetre waves it originates from the surrounding radiation belts, at decametre waves there are the amazing intense bursts; here the change of wavelength reveals quite different types of phenomena.

One of the extraordinary successes of radio methods has been the achievement of extremely high angular resolution, in certain circumstances comparable with and even surpassing that of optical telescopes. Contrary to early beliefs the attainment of sufficient resolution has not been an insurmountable obstacle in radio astronomy. Two simple factors largely explain why the anticipated fears proved groundless. The first is that radio signals received at widely-separated aerial positions can be linked together by cables, maintaining the correct phase and amplitude relation between them. Hence it becomes feasible to synthesise aerial apertures of huge size. Secondly, the very property that seemed such a detraction, the long length of radio waves, has turned out in certain respects to be a boon. Not only is it physically easier to make measurements involving dimensions of the order of metres or centimetres, but the factors causing variation of path length are less serious for radio since they represent a smaller fraction of the wavelength. A given path difference, measured in wavelengths is a million times smaller for radio than it is for light. Hence variations in path length, whether due to changes of dielectric constant or refraction or mechanical deformation, are far less serious for radio.

Radio astronomers are fully cognisant of the capabilities of electronics and can skilfully devise ingenious methods of achieving desired objectives. So, in addition to radio telescope systems interconnected by cables to synthesise large apertures, we now have interferometers operating over enormous baselines by utilising either radio links or independent recordings subsequently brought together for correlation. With the latter methods angular diameters of radio sources have been determined down to a thousandth of a second of arc, an achievement which would have been regarded twenty five years ago as utterly fantastic.

Lest it should be thought that I am implying that radio is acquiring a general superiority in resolving power in comparison with optical telescopes, I will elaborate certain limitations in image formation. Consider first the simple type of telescope with a single parabolic reflector. An optical telescope receives simultaneously and independently over many different directions. The photographic plate, or the retina of the eye, can be regarded as a multitude of receiving elements, each placed next to the other to observe in different directions and to form an optical image covering the field of view. A radio reflector, on the other hand, usually has a single receiving element at the focus to collect the radiation received over the radio beam, of width approximately λ/D radians. At radio wavelengths it has not proved practicable in single parabolic reflectors to set up multiple feeds to produce many beams simultaneously. Instead, mapping of radio emission has been achieved by scanning over the observed region with a single beam. The need to increase resolution led to the multielement systems effectively synthesising much larger apertures. At the same time, at the expense of added complexity, such systems made it feasible to interconnect the elements in different ways simultaneously so as to produce multiple beams (the Culgoora radio heliograph for instance produces 48 beams simultaneously). However, radio is still far from matching optical image-forming ability. The radio systems are subject to two restrictions. The multiple beam arrangements are complex. Also, the size of the synthesised aperture is limited to distances over which stable interconnection can be maintained, the maximum separation in present systems being of the order of 1 km. If we consider a maximum spacing of 1 km at a wavelength of 10 cm, the angular resolution is $\lambda/D = 1/10\,000$ radian. The 200 inch Palomar telescope, in comparison, with $\lambda = 0{\cdot}0005$ mm and $D = 5$ m, theoretically has a resolving power of $\lambda/D = 1/10\,000\,000$ radian excluding the effects of atmospheric turbulence. Of course, radio can call on the resources of very long baseline interferometry to pick out particular bright spots of radio emission with extremely high resolving power, surpassing that of optical instruments. We may infer that even if radio has not yet rivalled optical telescopes in image formation it has made enormous leaps towards the achievement of very high resolution.

Another interesting comparison arises from the question: in clear regions of the sky, which can probe furthest into the universe, radio or optical? If we compare the results achieved so far, it certainly appears that radio is capable of observing the most distant sources. The radio power received from Cygnus A for example is the second strongest of any source in the sky and yet it is optically very faint. Had it been at far greater range it would have been detectable only by radio. Many unidentified sources in radio surveys may be assumed to be sources too distant for optical detection even on long exposure photographs. The ultimate answer however depends both on the intrinsic radio and optical luminosity of astronomical sources, and on the technological perfection of receiving devices. In optics, for example, photoelectric detectors can greatly enhance reception sensitivity.

Let us consider the most luminous classes of sources at optical and radio wavelengths for they will provide our deepest look into the depths of the universe. If we compare the radio and optical luminosities of the strongest sources, such as the radio galaxy 3C 295, or the quasar 3C 147, we find remarkable similarity; both are of the order of 10^{45} ergs/sec. As a quantum of radio energy is about a million times less than that of light, there are about a million times more radio photons emitted by the source. Although this would seem to place radio reception at a great advantage, a compensating factor which helps to boost optical sensitivity is the wide bandwidth. The optical band covers altogether 4 to 8×10^{14} Hz that is, a bandwidth of about 4×10^{14} Hz, while the radio band comprises frequencies of up to about 10^{11} Hz. Actually, a normal bandwidth for a practical radio receiver rarely exceeds more than about 30 MHz, or 3×10^{7} Hz. Hence if we compare the bandwidth of a detector of white light with that of a radio receiver, the optical band is about 10^{7} times wider. It is clearly a matter of balancing factors. What is really important is that the sensitivities are so nearly comparable. When all is said on either side the question as to which method can reach the furthest distance tends to be academic. There are few types of analyses of distant sources that can depend on radio alone. Cosmological inferences from source counts is one particular example. For any precise radio investigation of distant sources, optical observations are an essential accompaniment in ascertaining both the distance and type of source. Optical and radio astronomy are complementary bringing quite different kinds of information about sources; both are essential to the realisation of any complete picture of astrophysics and of the structure and evolution of the universe.

It seems improbable that the situation will be radically altered by placing observatories on space vehicles, which are becoming a practical proposition. Freedom from atmospheric irregularities will greatly improve the attainable optical resolving power, and the darker sky background will increase sensitivity. For radio, the waveband will be extended by the freedom from atmospheric absorption at mm wavelengths, and from ionospheric reflection at the long wavelengths greater than 10 m. All these are valuable assets but the most spectacular advances in astronomical science derived from space observatories will almost certainly be in those parts of the spectrum of electromagnetic waves and fast particles that have previously been barred by the Earth's atmosphere. X-ray, ultraviolet and infrared astronomy are already becoming established.

Radio astronomy has grown from the combination of basic physics and radio technology applied to astronomy. It is the continual involvement with radio equipment and its performance that has in the past tended to set the radio astronomer in a separate category distinct from the optical astronomer. The necessary practical aspects of radio equipment can in fact be acquired through a comparatively short course of training. As radio astronomy has progressed, the physical aspects of problems encountered have asserted a more dominant role, the research worker becoming deeply concerned with the natural processes of production and propagation of radio waves, plasma physics, the influence of magnetic fields, diffraction theory, and the whole range of astrophysical and cosmological implications of his findings. Since 1945 the fusion with astronomy has proceeded gradually and comprehensively. While the radio astronomer has become more thoroughly immersed in the astronomical aims and significance of his research, at the same time the optical astronomers have continually helped to integrate the new discoveries and results with optical astronomy. An agreeable state of cooperation exists between radio and optical astronomers, maintaining their separate domains and observatories, yet frequently liaising and combining together whenever occasion demands. The interdependence that is now so essential has culminated from a natural growth of mutual interests and friendly cooperation.

As a research subject, radio astronomy presents great scope and challenge. In my opinion it offers an admirable training ground for a research graduate, whether he intends to continue in this field of research, or whether he subsequently plans a career elsewhere. The pursuit of radio astronomy combines a valuable range of disciplines. It encompasses branches of experimental and theoretical physics, astronomy, electronics and engineering, computing and statistical analysis; it involves the organisation and operation of systems, and working in collaboration with others is an essential requisite since large scale projects can only be tackled by teams.

After its hesitant beginning in the 1930s the science of radio astronomy has grown with astonishing rapidity. I am conscious that my outline of its development cannot portray more than a fraction of the totality of the research and achievement that has proceeded so intensively during the twenty five years 1945–70. The preparation of a book of economic length and reasonable readability precludes a more detailed account. In the Appendix, I mention just a few aspects which have not been discussed. What I have endeavoured to convey is an informative impression of the evolution and progress of one of the greatest advances in research of modern times.

It would be foolish to attempt to forecast future trends, for the essence of research is a venture into the unknown. It is within the bounds of possibility that some other civilisation might attempt to communicate with us by radio. Such startling eventualities we cannot predict. But we do know that a tremendous and impressive range of research has been opened up by radio astronomy. It is a young and flourishing science. We can be certain that radio methods in the future will have a major part in our quest for knowledge of the nature of the universe.

APPENDIX

Additional Aspects of Radio Astronomy Research

In an endeavour not to encumber the reader with too many topics I have deliberately omitted descriptions of certain aspects of research from the main text. I mention here a few such items as a guide to anyone wishing to seek further information.

Radio astronomical tests of the general theory of relativity:

The extraordinary precision attained in recent years has brought certain tests within the capabilities of radio techniques, in particular the influence of the solar gravitational field on the propagation of electromagnetic waves near to the Sun. High-resolution interferometry has made it possible to measure at centimetric wavelengths the gravitational deflection, of the order of a second of arc, of rays passing within close proximity of the Sun. The test has been successfully applied to the radio source 3C 279 as it approaches the direction of the Sun, the first results obtained at Cal. Tech. in 1969–70 demonstrating satisfactory agreement with theoretical expectations. With improving accuracy it will be possible to decide whether the deviations agree more closely with the Einstein predictions or with the Brans and Dicke theory. An alternative type of observational test initiated by Shapiro at Lincoln Lab. entailed measuring the gravitational delay, up to 200 μs, in the transmission of a radar pulse close to the Sun. Promising preliminary results have been achieved utilising radar echoes from Venus and Mercury at superior conjunction.

Zeeman splitting of the 21 cm hydrogen line:

The Zeeman splitting of spectral lines into polarised components in the presence of a magnetic field is well known in optics, and in 1907 Hale determined the field strength of sunspots by measuring the separation of the Zeeman components in optical line spectra. Bolton and Wild in 1957 suggested that the Zeeman splitting of the 21 cm radio line could provide a useful means of estimating galactic magnetic fields. Although the Zeeman splitting is small compared with the width of the line profiles, it has been shown, largely through the efforts of R. D. Davies and Verschuur at Jodrell Bank and at NRAO, that despite complexities in the spectra it has been proved feasible to derive data on the magnetic field in neutral hydrogen clouds by this method.

Interstellar radio communication:

It seems reasonable to suppose that intelligent beings exist elsewhere in the universe, for it is highly probable that many stars similar to the Sun have planetary systems and that a considerable proportion may possess a planet like the Earth with atmospheric and surface conditions able to support life. It is conceivable that certain planets may be populated by advanced civilisations who, believing that life could exist in our solar system, might be transmitting messages in the patient expectation of an eventual reply. Of course, time delays in transmission over astronomical distances present a severe handicap. Also, it is a remote chance that a technically competent civilisation should be within communicating distance and show sufficient concern and curiosity to embark on such an uncertain tentative exercise. It seems likely that radio would be employed for any attempted communication and, in consequence, the prospective detection of interstellar signals falls within the province of radio astronomy. An assessment of the problem was made in 1959 by Cocconi and Morrison who estimated that about 100 stars of the type which might possess habitable planets lie within a range of 50 light years. They believed a search for transmitted messages to be worthwhile, and suggested 1420 MHz as the most promising frequency since it would be known that radio astronomers anywhere in the universe would be likely to be recording at this hydrogen line frequency. Drake of NRAO thought the chances of success sufficiently within the bounds of possibility to warrant trial observations, and in 1960 he listened at frequencies near 1420 MHz at various times over a three-month period with an 85 ft radio telescope directed at two of the nearest candidates among the stars, Epsilon Eridani (10·8 light years away) and Tau Ceti (at 11·8 light years). The project was named Ozma after the story "The Wizard of Oz" about a far-away land of Oz peopled by strange beings. The Ozma experiment did not reveal any messages, but the attempt aroused interest. A book,[1] containing a collection of articles with reprints of papers written between 1959 and 1962 (including those by Drake), covers many aspects of the problem of possible life in the universe and intercommunication.

Although searching for interstellar messages may seem to be a tantalising pastime, few would question the profound significance of the detection of intelligent signals from other regions of the universe. The evidence of the nature of civilisations more advanced than our own would invoke implications of unprecedented philosophical and practical importance.

Other research topics:

The diversity of radio applications to astronomical problems precludes a comprehensive discussion. They include for instance the studies of neutral hydrogen in external galaxies, the attempts to determine the density of intergalactic hydrogen, the investigation of the magnetic field of the Galaxy from the pattern of Faraday rotation in the linearly-polarised components of distant radio sources, and so on. It is interesting to note that although the original intention at Jodrell Bank of obtaining radar echoes from cosmic ray showers was not fulfilled, bursts of radio emission by Cerenkov radiation from cosmic ray showers entering the atmosphere were detected in an experimental investigation in 1965 at Jodrell Bank by a team led by F. G. Smith and J. V. Jelley. The notes on a few further examples of radio astronomical research outlined in this Appendix suffice to emphasise the great versatility of radio methods.

[1] Interstellar Communication. Ed. A. G. W. Cameron (Benjamin, 1963).

References

1932

JANSKY, K. G., "Directional studies of atmospherics at high frequencies", *Proc. IRE*, **20**, 1920.

SCHAFER, J. P., and W. M. GOODALL, "Observations of Kennelly-Heaviside layer heights during the Leonid meteor shower, November 1931", *Proc. IRE*, **20**, 1941.

SKELLETT, A. M., "The ionising effect of meteors in relation to radio propagation", *Proc. IRE*, **20**, 1933.

1933

JANSKY, K. G., "Electrical disturbances apparently of extraterrestrial origin", *Proc. IRE*, **21**, 1387.

JANSKY, K. G., "Radio waves from outside the solar system", *Nature*, **132**, 66.

1935

JANSKY, K. G., "A note on the source of interstellar interference", *Proc. IRE*, **23**, 1158.

1938

PIERCE, J. A., "Abnormal ionisation in the E region of the ionosphere", *Proc. IRE*, **26**, 892.

1940

HENYEY, L. G., and P. C. KEENAN, "Interstellar radiation from free electrons and hydrogen atoms", *Ap. J.*, **91**, 625.

REBER, G., "Cosmic static", *Proc. IRE*, **28**, 68.

REBER, G., "Cosmic static", *Ap. J.*, **91**, 621.

1941

BLACKETT, P. M. S., and A. C. B. LOVELL, "Radio echoes and cosmic ray showers", *Proc. Roy. Soc.*, **A 177**, 183.

CHAMANLAL and K. VENKATARAMAN, "Whistling meteors—a Doppler effect produced by meteors entering the ionosphere", *Electrotechnics*, **14**, 28.

PIERCE, J. A., "A note on ionisation by meteors", *Phys. Rev.*, **59**, 625.

1942

HEY, J. S., "Metre-wave radiation from the Sun". See 1946. *AORG Report (Restricted Circulation)*.

SOUTHWORTH, G. C., "Microwave radiation from the Sun". See 1945. *BTL Report (Restricted Circulation)*.

1944

REBER, G., "Cosmic static", *Ap. J.*, **100,** 279.

1945

APPLETON, E. V., "Departure of long-wave solar radiation from black-body intensity", *Nature*, **156,** 534.

SOUTHWORTH, G. C., "Microwave radiation from the Sun", *J. Franklin Inst.*, **239,** 285.

VAN DE HULST, H. C., "Radio waves from space", *Ned. Tij. Natuurk.*, **11,** 201, 210.

1946

APPLETON, E. V., and J. S. HEY, "Solar radio noise", *Phil. Mag.*, **37,** 73.

ARTSIMOVICH, L. A., and I. YA. POMERANCHUK, "The radiation of fast electrons in a magnetic field", *Zh. Eksp. Terr. Fiz.*, **16,** 379.

BAY, Z., "Reflections of microwaves from the Moon", *Hung. Acta Phys.*, **1,** 1.

DICKE, R. H., 'The measurement of thermal radiation at microwave frequencies", *Rev. Sci. Instr.*, **17,** 268.

DICKE, R. H., and R. BERINGER, "Microwave radiation from the Sun and Moon", *Ap. J.*, **103,** 375.

DICKE, R. H., R. BERINGER, R. L. KYHL, A. B. VANE, "Atmospheric absorption measurements with a microwave radiometer", *Phys. Rev.*, **70,** 340.

FORBUSH, S. E., "Three unusual cosmic ray increases possibly due to charged particles from the Sun", *Phys. Rev.*, **70,** 771.

GAMOW, G., "Expanding universe and the origin of the elements", *Phys. Rev.*, **70,** 572.

GINZBURG, V. L., "On solar radiation in the radio spectrum", *C. R. Acad. Sci. USSR*, **52,** 487.

HEY, J. S., "Solar radiations in the 4–6 metre radio wavelength band", *Nature*, **157,** 47.

HEY, J. S., S. J. PARSONS, J. W. PHILLIPS, "Fluctuations in cosmic radiation at radio frequencies", *Nature*, **158,** 234.

HEY, J. S., and G. S. STEWART, "Derivation of meteor stream radiants by radio reflection methods", *Nature*, **158,** 481.

MARTYN, D. F., "Temperature radiation from the quiet Sun in the radio spectrum", *Nature*, **158,** 632.

MOFENSON, J., "Radio echoes from the Moon", *Electronics*, **19,** 92.

PAWSEY, J. L., "Observation of million degree radiation from the Sun at a wavelength of 1·5 metres", *Nature*, **158,** 633.

RYLE, M., and D. D. VONBERG, "Solar radiation on 175 Mc/s", *Nature*, **158,** 339.

SHKLOVSKY, I. S., 'On the radiation of radio waves by the Galaxy and by upper layers of the solar atmosphere", *Astron. Zh.*, **23,** 333.

WEBB, H. D., "Project Diana—Army radar contacts the Moon", *Sky and Telescope*, **5,** 3.

1947

APPLETON, E. V., and R. NAISMITH, "The radio detection of meteor trails and allied phenomena", *Proc. Phys. Soc.*, **59,** 461.

HEY, J. S., and G. S. STEWART, "Radar observation of meteors", *Proc. Phys. Soc.*, **59,** 858.

HEY, J. S., S. J. PARSONS, G. S. STEWART, "Radar observations of the Giacobinid meteor shower", *MNRAS*, **107,** 176.

KHAIKIN, S. E., and B. M. CHIKHACHEV, "Investigations of radio emission from the Sun . . . during the solar eclipse of 20 May, 1947", *C. R. Acad. Sci., USSR*, **58,** 1923.

LOVELL, A. C. B., C. J. BANWELL, J. A. CLEGG, "Radio echo observations of the Giacobinid meteors, 1946", *MNRAS*, **107,** 164.

MCCREADY, L. L., J. L. PAWSEY, R. PAYNE-SCOTT, "Solar radiation at radio frequencies and its relation to sunspots", *Proc. Roy. Soc.*, **A190,** 357.

PAYNE-SCOTT, R., D. E. YABSLEY, J. G. BOLTON, "Relative times of arrival of bursts of solar noise on different radio frequencies", *Nature*, **160,** 256.

1948

BOLTON, J. G., and G. J. STANLEY, "Observations on the variable source of cosmic radio frequency radiation in the constellation of Cygnus", *Aust. J. Sci. Res.*, **A1,** 58.

CLEGG, J. A., "Determination of meteor radiants by observation of radio echoes from meteor trails", *Phil. Mag.*, **39,** 577.

COVINGTON, A. E., "Solar noise observations on 10·7 cm", *Proc. IRE*, **36,** 454.

ELLYETT, C. D., and J. G. DAVIES, "Velocity of meteors measured by diffraction of radio waves from trails during formation", *Nature*, **161,** 596.

HERLOFSON, N., "The theory of meteor ionisation", *Rep. Prog. Phys.*, **11,** 444.

HEY, J. S., J. W. PHILLIPS, S. J. PARSONS, "An investigation of galactic radiation in the radio spectrum", *Proc. Roy. Soc.*, **A 192,** 425.

IVANENKO, D. D., and A. A. SOKOLOV, "On the theory of the 'luminous' electron", *Dokl. Akad. Nauk. SSSR*, **59,** 1551.

MARTYN, D. F., "Solar radiation in the radio spectrum", *Proc. Roy. Soc.*, **A 193,** 44.

RYLE, M., and F. G. SMITH, "A new intense source of radio-frequency radiation in the constellation of Cassiopeia", *Nature*, **162,** 462.

RYLE, M., and D. D. VONBERG, "An investigation of radio frequency radiation from the Sun", *Proc. Roy. Soc.*, **A 193,** 98.

1949

ALPHER, R. A., and R. C. HERMAN, "Remarks on the evolution of the expanding universe", *Phys. Rev.*, **75,** 1089.

BOLTON, J. G., and G. J. STANLEY, "The position and probable identification of the source of galactic radio-frequency radiation Taurus A", *Aust. J. Sci. Res.*, **A 2,** 139.

BOLTON, J. G., G. J. STANLEY, O. B. SLEE, "Positions of three discrete sources of galactic radio-frequency radiation", *Nature*, **164,** 101.

CHRISTIANSEN, W. N., D. E. YABSLEY, B. Y. MILLS, "Measurements of solar radiation at a wavelength of 50 cm during the eclipse of 1 November 1948", *Aust. J. Sci. Res.*, **A 2,** 506.

COVINGTON, A. E., "Circularly polarised solar radiation on 10·7 cm", *Proc. IRE*, **37**, 407.

DE WITT, J. H., and E. K. STODOLA, "Detection of radio signals reflected from the Moon", *Proc. IRE*, **37**, 229.

MANNING, L. A., O. G. VILLARD, A. M. PETERSEN, "Radio Doppler investigation of meteoric heights and velocities", *J. Appl. Phys.*, **20**, 475.

PIDDINGTON, J. H., and H. C. MINNETT, "Microwave thermal radiation from the Moon", *Aust. J. Sci. Res.*, **A 2**, 63.

SCHWINGER, J., "On the classical radiation from accelerated electrons", *Phys. Rev.*, **75**, 912.

1950

ALFVÉN, H., and N. HERLOFSON, "Cosmic radiation and radio stars", *Phys. Rev.*, **78**, 616.

GETMANZEV, G. G., and GINZBURG, V. L., "On the diffraction of solar and cosmic radio emission by the Moon", *Zh. Eksper. Terr. Fiz.*, **20**, 347.

KIEPENHEUER, K. O., "Cosmic rays as the source of general galactic radiation", *Phys. Rev.*, **79**, 738.

RYLE, M., F. G. SMITH, B. ELSMORE, "Preliminary survey of the radio stars in the northern hemisphere", *MNRAS*, **110**, 508.

SMERD, S. F., "Radio frequency radiation from the quiet Sun", *Aust. J. Sci. Res.*, **A 3**, 34.

SMITH, F. G., C. G. LITTLE, A. C. B. LOVELL, "Origin of the fluctuations in the intensity of radio waves from galactic sources", *Nature*, **165**, 422.

STANIER, H. M., "Distribution of radiation from the undisturbed Sun at a wavelength of 60 cm", *Nature*, **165**, 354.

WALDMEIER, M., and H. MÜLLER, "Solar radiation in the region of $\lambda =$ 10 cm", *Z. Astr.*, **27**, 58.

WILD, J. P., and L. L. MCCREADY, "Observations of the spectrum of high intensity solar radiation at metre wavelengths. Pt I The apparatus and spectral types", *Aust. J. Sci. Res.*, **A 3**, 387.

WILD, J. P., "Observations of the spectrum of high intensity solar radiation at metre wavelengths. Pt II Outbursts. Pt III Isolated bursts", *Aust. J. Sci. Res.*, **A 3**, 399, 541.

1951

ALMOND, M., J. G. DAVIES, A. C. B. LOVELL, "The velocity distribution of sporadic meteors", *MNRAS*, **111**, 585; **112**, 21 (1952); **113**, 411 (1953).

BROWN, R. H., and C. HAZARD, "Radio emission from the Andromeda Nebula", *MNRAS*, **111**, 357.

EWEN, H. I., and E. M. PURCELL, "Radiation from galactic hydrogen at 1420 Mc/s", *Nature*, **168**, 356.

HAGEN, J. P., "Temperature gradient in the Sun's atmosphere measured at radio frequencies", *Ap. J.*, **113**, 547.

KERR, F. J., and C. A. SHAIN, "Moon echoes and transmission through the ionosphere", *Proc. IRE*, **39**, 230.

LITTLE, A. G., and R. PAYNE-SCOTT, "The position and movement on the solar disk of sources of radiation at a frequency of 97 Mc/s", *Aust. J. Sci. Res.*, **A 4**, 489.

MACHIN, K. E., and F. G. SMITH, "A new method for measuring the electron density in the solar corona", *Nature*, **168,** 599.

MCKINLEY, D. W. R., "Meteor velocities determined by radio observations", *Ap. J.*, **113,** 225.

MULLER, C. A., and J. H. OORT, "The interstellar hydrogen line at 1420 Mc/s and an estimate of galactic rotation", *Nature*, **168,** 357.

SMITH, F. G., "An accurate determination of the positions of four radio stars", *Nature*, **168,** 555.

VITKEVICH, V. V. "A new method for investigating the solar corona", *Dokl. Akad. Nauk. SSSR*, **77,** 585.

1952

BROWN, R. H., and C. HAZARD, "Radio frequency radiation from Tycho Brahe's supernova", *Nature*, **170,** 364.

BROWN, R. H., R. C. JENNISON, M. K. DAS GUPTA, "Apparent angular sizes of discrete radio sources", *Nature*, **170,** 1061.

CHRISTIANSEN, W. N., and J. V. HINDMAN, "A preliminary survey of 1420 Mc/s line emission from galactic hydrogen", *Aust. J. Sci. Res.*, **A 5,** 437.

KERR, F. J., "On the possibility of obtaining radar echoes from the Sun and planets", *Proc. IRE*, **40,** 660.

MACHIN, K. E., and F. G. SMITH, "Occultation of a radio star by the solar corona", *Nature*, **170,** 319.

MILLS, B. Y., "Apparent angular sizes of discrete radio sources", *Nature*, **170,** 1063.

PAYNE-SCOTT, R., and A. G. LITTLE, "The position and movement on the solar disk of sources of radiation at a frequency of 97 Mc/s. Pt III Outbursts", *Aust. J. Sci. Res.*, **A 5,** 32.

RYLE, M., "A new radio interferometer and its application to the observation of weak radio sources", *Proc. Roy. Soc.*, **A 211,** 351.

SHKLOVSKY, I. S., "On the nature of radio emission from the Galaxy", *Astron. Zh.*, **29,** 418.

VITKEVICH, V. V., "Interferometer methods for radio astronomy", *Astron. Zh.*, **29,** 450.

1953

CHRISTIANSEN, W. N., and J. A. WARBURTON, "The distribution of radio brightness over the solar disk at a wavelength of 21 cm", *Aust. J. Phys.*, **6,** 262.

JENNISON, R. C., and M. K. DAS GUPTA, "Fine structure of the extraterrestrial source Cygnus", *Nature*, **172,** 996.

MILLS, B. Y., "The radio brightness distribution over four discrete sources of cosmic noise", *Aust. J. Phys.*, **6,** 452.

MILLS, B. Y., and A. G. LITTLE, "A high resolution aerial system of a new type", *Aust. J. Phys.*, **6,** 272.

SHLOVSKY, I. S., "On the nature of the radiation from the Crab Nebula", *Dokl. Akad. Nauk. SSSR*, **90,** 983.

SHKLOVSKY, I. S., "The possibility of observing monochromatic radio emissions from interstellar molecules", *Dokl. Akad. Nauk. SSSR*, **92,** 25.

WILD, J. P., J. D. MURRAY, W. C. ROWE, "Evidence of harmonics in the spectrum of a solar radio burst", *Nature*, **172,** 533.

1954

BAADE, W., and R. MINKOWSKI, "Identification of the radio sources in Cassiopeia, Cygnus A, and Puppis A", *Ap. J.*, **119,** 206.

BRACEWELL, R. N., and J. A. ROBERTS, "Aerial smoothing in radio astronomy", *Aust. J. Phys.*, **7,** 615.

BROWN, R. H., and R. Q. TWISS, "A new type of interferometer for use in radio astronomy", *Phil. Mag.*, **45,** 663.

DOMBROVSKY, V. A., "On the nature of the radiation from the Crab Nebula", *Dokl. Akad. Nauk. SSSR.*, **94,** 1021.

HADDOCK, F. T., C. H. MAYER, R. M. SLOANAKER, "Radio emission from the Orion Nebula and other sources at $\lambda = 9{\cdot}4$ cm", *Ap. J.*, **199,** 456.

HAGEN, J. P. and E. F. MCCLAIN, "Galactic absorption of radio waves", *Ap. J.*, **120,** 368.

KERR, F. J., J. V. HINDMAN, B. J. ROBINSON, "Observations of the 21 cm line from the Magellanic Clouds", *Aust. J. Phys.*, **7,** 297.

MURRAY, W. A. S., and J. K. HARGREAVES, "Lunar radar echoes and the Faraday effect in the ionosphere", *Nature*, **173,** 944.

TROITSKY, V. S., "A contribution to the theory of the radio emission from the Moon", *Astron. Zh.*, **31,** 511.

VAN DE HULST, H. C., C. A. MULLER, J. H. OORT, "The spiral structure of the outer part of the galactic system derived from the hydrogen emission at 21 cm wavelength", *BAN*, **12,** 117.

WILD, J. P., J. D. MURRAY, W. C. ROWE, "Harmonics in the spectra of solar radio disturbances", *Aust. J. Phys.*, **7,** 439.

WILLIAMS, D. R. W., and R. D. DAVIES, "A method for the measurement of the distance of radio sources", *Nature*, **173,** 1182.

1955

BALDWIN, J. E., "(a) A survey of the integrated radio emission at a wavelength of $3{\cdot}7$ m. (b) The distribution of galactic radio emission", *MNRAS*, **115**, (a) 684, (b) 690.

BROWN, R. H., H. P. PALMER, A. R. THOMPSON, "A rotating lobe interferometer and its application to radio astronomy", *Phil. Mag.*, **46,** 857.

BURKE, B. F., and K. L. FRANKLIN, "Observations of a variable radio source associated with the planet Jupiter", *J. Geophys. Res.*, **60,** 213.

HAGEN, J. P., A. E. LILLEY, E. F. MCCLAIN, "Absorption of 21 cm radiation by interstellar hydrogen", *Ap. J.*, **122,** 361.

HEWISH, A., "The irregular structure of the outer regions of the solar corona", *Proc. Roy. Soc.*, **A 228,** 238.

MILLS, B. Y., "The observation and interpretation of radio emission from some bright galaxies", *Aust. J. Phys.*, **8,** 368.

SHAKESHAFT, J. R., M. RYLE, J. E. BALDWIN, B. ELSMORE, J. H. THOMSON, "A survey of radio sources between declinations $-38°$ and $+83°$", *Mem. RAS*, **67,** 106.

VITKEVICH, V. V., "Results of observations of scattering of radio waves through the solar corona", *Astron. Zh.*, **32,** 150.

1956

BLOEMBERGEN, N., "Proposal for a new type of solid state maser", *Phys. Rev.*, **104,** 324.

BROWN, R. H., and R. Q. TWISS, "A test of a new type of stellar interferometer on Sirius", *Nature*, **178,** 1046.

BROWNE, I. C., J. V. EVANS, J. K. HARGREAVES, W. A. S. MURRAY, "Radio echoes from the Moon", *Proc. Phys. Soc.*, **B 69,** 901.

BURBIDGE, G. R., "On the synchrotron radiation from M87", *Ap. J.*, **124,** 416.

DRÖGE, F., and W. PRIESTER, "Distribution of galactic radiation at 200 MHz", *Z. Astrophys.*, **40,** 236.

GILL, J. C., and J. G. DAVIES, "A radio echo method of meteor orbit determination", *MNRAS*, **116,** 105.

GINZBURG, V. L., "On non-ionospheric fluctuations of the intensity of radio waves from nebulae", *Sov. Phys. Dokl.*, **1,** 403.

SHAIN, C. A., "18·3 Mc/s radiation from Jupiter", *Aust. J. Phys.*, **9,** 61.

1957

BLYTHE, J. H., "(a) A new type of pencil beam for radio astronomy. (b) Results of a survey of galactic radiation at 38 Mc/s", *MNRAS*, **117,** (a) 644, (b) 652.

BOLTON, J. G., and J. P. WILD, "On the possibility of measuring interstellar magnetic fields by 21 cm Zeeman splitting", *Ap. J.*, **125,** 296.

EVANS, J. V., "The scattering of radio waves by the Moon", *Proc. Phys. Soc.*, **B 70,** 1105.

MAYER, C. H., T. P. McCULLOUGH, R. M. SLOANAKER, "Evidence for polarised radiation from the Crab Nebula", *Ap. J.*, **126,** 468.

MULLER, C. A., and G. WESTERHOUT, "A catalogue of 21 cm line profiles", *BAN*, **13,** 201.

SCHMIDT, M., "Spiral structure in the inner parts of the galactic system derived from the hydrogen emission at 21 cm wavelength", *BAN*, **13,** 247.

TOWNES, C. H., "Microwave and radio frequency resonance lines of interest to radio astronomy", *Radio Astronomy*, *IAU Symp.* **No 4,** 92.

1958

BOISCHOT, A., "Study of solar radio emission on 169 MHz with a large interferometer system", *Ann. Astrophys.*, **21,** 273.

BRACEWELL, R. N., "Radio interferometry of discrete sources", *Proc. IRE*, **46,** 97.

GIBSON, J. E., "Lunar thermal radiation at 35 KMC", *Proc. IRE*, **46,** 280.

HEWISH, A., "The scattering of radio waves in the solar corona", *MNRAS*, **118,** 534.

MAXWELL, A., and G. SWARUP, "A new spectral characteristic in solar radio emission", *Nature*, **181,** 36.

MAYER, C. H., T. P. McCULLOUGH, R. M. SLOANAKER, "(a) Observations of Venus at 3·15 cm wavelength. (b) Observations of Mars and Jupiter at a wavelength of 3·15 cm", *Ap. J.*, **127,** (a)1, (b)11.

MILLS, B. Y., O. B. SLEE, E. R. HILL, "A catalogue of radio sources between declinations +10° −30°", *Aust. J. Phys.*, **11**, 360.

OORT, J. H., F. J. KERR, G. WESTERHOUT, "The galactic system as a spiral nebula", *MNRAS*, **118**, 379.

RYLE, M., "The nature of cosmic radio sources", *Proc. Roy. Soc.*, **A 248**, 289.

WESTERHOUT, G., "A survey of the continuous radiation from the galactic system at a frequency of 1390 Mc/s", *BAN*, **14**, 215.

1959

BROWN, R. H., and C. HAZARD, "The radio emission from normal galaxies" *MNRAS*, **119**, 297.

COCCONI, G., and P. MORRISON, "Searching for interstellar communications", *Nature*, **184**, 844.

DRAKE, F. D., and H. HVATUM, "Non-thermal microwave radiation from Jupiter", *Astron. J.*, **64**, 329.

EDGE, D. O., J. R. SHAKESHAFT, W. E. MCADAM, J. E. BALDWIN, S. ARCHER, "A survey of radio sources at a frequency of 159 Mc/s (3C Survey)", *Mem. RAS*, **68**, 37.

GIBSON, J. E., and R. J. McEWAN, "Observations of Venus at 8·6 mm wavelength", *Paris Symp. Radio Astron*, **50.**

GINZBURG, V. L., and ZHELEZNYAKOV, V. V., "On the propagation of electromagnetic waves in the solar corona taking into account the influence of the magnetic field", *Sov. Astron. AJ*, **3**, 235.

GIORDMAINE, J. A., L. E. ALSOP, C. H. MAYER, C. H. TOWNES, "A maser amplifier for radio astronomy at X-band", *Proc. IRE*, **47**, 1062.

KARDASHEV, N. S., "On the possibility of detection of allowed lines of atomic hydrogen in the radio frequency spectrum", *Sov. Astron. AJ*, **3**, 813.

KERR, F. J., J. V. HINDMAN, C. S. GUM, "A 21 cm survey of the Southern Milky Way", *Aust. J. Phys.*, **12**, 270.

KUNDU, M. R., "Structure and properties of sources of solar activity at centimetric waves", *Ann. Astrophys.*, **22**, 1.

MILLS, B. Y., "The radio continuum from the Galaxy", *PASP*, **71**, 267.

ROBERTS, J. A., "Solar radio bursts of spectral type II", *Aust. J. Phys.*, **12**, 327.

SLOANAKER, R. M., "Apparent temperature of Jupiter at a wavelength of 10 cm", *Astron. J.*, **64**, 346.

WILD, J. P., K. V. SHERIDAN, A. A. NEYLAN, "An investigation of the speed of the solar disturbances responsible for Type III radio bursts", *Aust. J. Phys.*, **12**, 369.

1960

BLAAUW, A., C. S. GUM, F. J. KERR, J. H. OORT, J. L. PAWSEY, G. W. ROUGOOR, G. WESTERHOUT, "The new IAU system of galactic coordinates" (5 papers). *MNRAS*, **121**, 123.

ESHLEMAN, V. R., R. C. BARTHLE, P. B. GALLAGHER, "Radar echoes from the Sun", *Science*, **131**, 329.

HÖGBOM, J. A., "The structure and magnetic field of the solar corona", *MNRAS*, **120,** 530.

PARKER, E. N., "The hydrodynamic theory of solar corpuscular radiation and stellar winds", *Ap. J.*, **132,** 821.

PETTENGILL, G. H., "Measurements of lunar reflectivity using the Millstone radar", *Proc. IRE*, **48,** 933.

RADHAKRISHNAN, V., and J. A. ROBERTS, "Polarisation and angular extent of the 960 Mc/s radiation from Jupiter", *Phys. Rev.*, **4,** L493.

RYLE, M., and A. HEWISH, "The synthesis of large radio telescopes", *MNRAS*, **120,** 220.

ROUGOOR, G. W., and J. H. OORT, "Distribution and motion of interstellar hydrogen in the galactic system with particular reference to the region within 3 kpc of the centre", *Proc. Nat. Acad. Sci.*, **46,** 1.

SHKLOVSKY, I. S., "Secular variations of the flux and intensity of radio emission from discrete sources", *Astron Zh.* **37**, 256; *Sov. Astron, AJ*, **4,** 243.

SHKLOVSKY, I. S., "Radio galaxies", *Astron Zh.* **37,** 945, *Sov. Astron, AJ*, **4,** 885.

VITKEVICH, V. V., "The solar corona from the observations of 1951–1958", *Sov. Astron. AJ*, **4,** 31.

1961

BENNETT, A. S., "The revised 3C catalogue of radio sources", *Mem. RAS*, **68,** 163.

BROWN, R. H., and C. HAZARD, "The radio emission from normal galaxies", *MNRAS*, **122,** 479; **123,** 279.

BURBIDGE, G. R., "Galactic explosions as sources of radio emission", *Nature*, **190,** 1053.

COATES, R. J., "Lunar brightness variations with phase at 4·3 mm wavelength", *Ap. J.*, **133,** 723.

GINZBURG, V. L., and V. V. ZHELEZNYAKOV, "Non-coherent mechanisms of sporadic solar radio emission in the case of a magneto-active coronal plasma", *Sov. Astron. AJ*, **5,** 1.

HAZARD, C., "Lunar occultations of a radio source", *Nature*, **191,** 58.

HÖGBOM, J. A., and J. R. SHAKESHAFT, "Secular variation of the flux density of the radio source Cassiopeia A", *Nature*, **189,** 561.

KUNDU, M. R., and F. T. HADDOCK, "Centimeter wave solar bursts and associated effects", *IRE Trans. AP*, **9,** 82.

LE ROUX, E., "Theoretical study of synchrotron radiation from radio sources", *Ann. Astrophys.*, **24,** 71.

SLEE, O. B., "Observations of the solar corona out to 100 solar radii", *MNRAS*, **123,** 223.

Radar observations of Venus in 1961: the references are—

MARON, I., *et al.*, *Science*, **134,** 1419 (1961).

THOMSON, J. H., *et al.*, *Nature*, **190,** 519 (1961).

KOTELNIKOV, V. A., *et al.*, *Radiotech. Electron.*, **7,** 1860 (1962).

MUHLEMAN, D. O., *et al.*, *Astron. J.*, **67,** 191 (1962).

PETTENGILL, G. H., *et al.*, *Astron. J.*, **67,** 181 (1962).

1962

ALLEN, L. R., R. H. BROWN, H. P. PALMER, "An analysis of the angular size of radio sources", *MNRAS*, **125,** 57.

ANDERSON, B., H. P. PALMER, B. ROWSON, "Brightness distribution of the radio source 14N5A", *Nature*, **195,** 165.

BRACEWELL, R. N., B. F. C. COOPER, T. E. COUSINS, "Polarisation in the central component of Centaurus A", *Nature*, **195**, 1289.

COOPER, B. F. C., and R. M. PRICE, "Faraday rotation effects associated with the radio source Centaurus A", *Nature*, **195,** 1084.

HOWARD, W. E., A. H. BARRETT, F. T. HADDOCK, "Measurements of the microwave radiation from the planet Mercury", *Ap. J.*, **136,** 995.

KOTELNIKOV, V. A., *et al.*, "Radar location of the planet Mercury", *Dokl. Akad. Nauk. SSSR*, **148,** 1320.

LARGE, M. I., M. J. S. QUIGLEY, C. G. T. HASLAM, 'A new feature of the radio sky", *MNRAS*, **124,** 405.

MALTBY, P., and A. T. MOFFET, 'Brightness distribution in discrete radio sources", *Ap. J. Suppl.*, **7,** 141.

MAYER, C. H., T. P. MCCULLOUGH, R. M. SLOANAKER, 'Evidence for polarised 3·15 cm radiation from the radio galaxy Cygnus A", *Ap. J.*, **135,** 656.

MORRIS, D., and G. L. BERGE, "Measurements of the polarisation and extent of the decimeter radiation from Jupiter", *Ap. J.*, **136,** 276.

MENON, T. K., "(a) A study of the Rosette nebula. (b) Physical conditions in the Orion nebula", *Ap. J*, (a) **135,** 394, (b) **136,** 95.

SALOMONOVICH, A. E., "Lunar radio emission in the cm band and some characteristics of the surface layer", *Sov. Astron. AJ*, **6,** 55.

SALOMONOVICH, A. E., "Solar radio emission on a wavelength of 8 mm", *Sov. Astron. AJ*, **6,** 202.

SCHEUER, P. A. G., "On the use of lunar occultations for investigating the angular structure of radio sources", *Aust. J. Phys.*, **15,** 333.

SMERD, S. F., J. P. WILD, K. V. SHERIDAN, "On the relative position and origin of harmonics in the spectra of solar radio bursts of spectral types II and III", *Aust. J. Phys.*, **15,** 180.

TAKAKURA, T., "Solar radio outbursts and the acceleration of electrons", *J. Phys. Soc. Japan*, **17,** Suppl. AII, 243.

WESTERHOUT, G., C. L. SEEGER, W. N. BROUW, J. TINBERGEN, 'Polarisation of galactic 75 cm radiation", *BAN*, **16,** 187.

1963

BURBIDGE, G. R., E. M. BURBIDGE, A. SANDAGE, "Evidence of violent events in the nuclei of galaxies", *Rev. Mod. Phys.*, **35,** 947.

CONWAY, R. G., K. I. KELLERMANN, R. J. LONG, "The radio frequency spectra of discrete radio sources", *MNRAS*, **125,** 261.

EVANS, J. V., and G. H. PETTENGILL, "The scattering behaviour of the Moon at wavelengths of 3·6, 68 and 784 cm", *J. Geophys. Res.*, **68,** 423.

GARDNER, F. F., and J. B. WHITEOAK, "Polarisation of radio sources and Faraday rotation effects in the Galaxy", *Nature*, **197,** 1162.

GREENSTEIN, J. L., T. A. MATTHEWS, "Redshift of the unusual radio source 3C, 48", *Nature*, **197,** 1041.

HAZARD, C., M. B. MACKEY, A. J. SHIMMINS, "Investigation of the radio source 3C 273 by the method of lunar occultations", *Nature*, **197,** 1037.

HEWISH, A., J. D. WYNDHAM, "The solar corona in interplanetary space", *MNRAS*, **126,** 469.

HOYLE, F., W. A. FOWLER, "On the nature of strong radio sources", *Nature*, **197,** 533; *MNRAS*, **125,** 169.

LOVELL, A. C. B., F. L. WHIPPLE, L. H. SOLOMON, "Radio emission from flare stars", *Nature*, **198,** 229.

MATTHEWS, T. A., A. R. SANDAGE, "Optical identification of 3C 48, 3C 196, 3C 286 with stellar objects", *Ap. J.*, **138,** 30.

ROWSON, B., "High resolution observations with a tracking interferometer", *MNRAS*, **125,** 177.

SCHMIDT, M., "3C 273: a star-like object with a large redshift", *Nature*, **197,** 1040.

SLEE, O. B., L. H. SOLOMON, G. E. PATSON, "Radio emission from the flare star V 371 Orionis", *Nature*, **199,** 991.

SLISH, V. I., "Angular size of radio sources", *Nature*, **199,** 682.

SMITH, H. J., D. HOFFLEIT, "Light variability and the nature of 3C 273", *Astron. J.*, **68,** 292.

SOBOLEVA, N. S., "Measurements of the polarisation of lunar radio emission on a wavelength of 3·2 centimetres", *Sov. Astron. AJ.*, **6,** 873.

WEINREB, S., A. H. BARRETT, M. L. MEEKS, J. C. HENRY, "Radio observations of OH in the interstellar medium", *Nature*, **200,** 829.

WILLIAMS, P. J. S., "Absorption of radio sources of high brightness temperature", *Nature*, **200,** 56.

1964

ADGIE, R. L., "Comparison of radio and optical positions of some identified radio sources", *Nature*, **204,** 1028.

BIGG, E. K., "Influence of the satellite Io on Jupiter's decametric radiation", *Nature*, **203,** 1008.

BOLTON, J. G., K. J. VAN DAMME, F. F. GARDNER, B. J. ROBINSON, "Observation of OH absorption lines in the radio spectrum of the galactic centre", *Nature*, **201,** 279.

CARPENTER, R. L., "Study of Venus by CW radar", *Astron. J.*, **69,** 2.

DRAVSKIKH, Z. V., A. F. DRAVSKIKH, V. A. KOLBASON, "An excited hydrogen line profile in the Omega nebula", *Astr. Tsirk.*, No 305, 2.

GOLDSTEIN, R. N., "Venus characteristics by Earth-based radar", *Astron. J.*, **69,** 12.

HEESCHEN, D. S., C. M. WADE, "A radio survey of galaxies", *Astron. J.*, **69,** 277.

HEWISH, A., P. F. SCOTT, D. WILLS, "Interplanetary scintillations of small diameter radio sources", *Nature*, **203,** 1214.

JAMES, J. C., "Radar echoes from the Sun", *IEEE Trans. Mil.*, **8,** 210.

MATTHEWS, T. A., W. W. MORGAN, M. SCHMIDT, "A discussion of galaxies identified with radio sources", *Ap. J.*, **140,** 35.

MATHEWSON, D. S., D. K. MILNE, "A pattern in the large-scale distribution of galactic polarised radio emission", *Nature*, **203**, 1273.

ROBINSON, B. J., F. F. GARDNER, K. J. VAN DAMME, J. G. BOLTON, "An intense concentration of OH near the galactic centre", *Nature*, **202**, 989.

SLEE, O. B., C. S. HIGGINS, "The apparent sizes of the Jovian decametric radio sources", *Aust. J. Phys.*, **19**, 167.

1965

ADGIE, R. L., H. GENT, O. B. SLEE, A. D. FROST, H. P. PALMER, B. ROWSON, "New limits to the angular sizes of some quasars", *Nature*, **208**, 275.

CLARK, B. G., A. D. KUZ'MIN, "The measurement of the polarisation and brightness distribution of Venus at 10.6 cm wavelength", *Ap.J.*, **142**, 23.

COOPER, B. F. C., R. M. PRICE, D. J. COLE, "A study of the decametric emission and polarisation of Centaurus A", *Aust. J. Phys.*, **18**, 589.

DENT, W. A., "Quasi-stellar sources: variation in the radio emission of 3C 273", *Science*, **148**, 1458.

DICKE, R. H., P. J. E. PEEBLES, P. G. ROLL, D. T. WILKINSON, "Cosmic black-body radiation", *Ap. J.*, **142**, 414.

EPSTEIN, E. E., "(a) 3·4 mm observations of quasi-stellar radio source 3C279. (b) Preliminary results on variations in the 3·4 mm flux from 3C 273", *Ap. J.*, **142**, (a) 1282, (b) 1285.

LOW, F. J., "Observations of 3C 273 and 3C 279 at 1 mm", *Ap. J.*, **142**, 1287.

PENZIAS, A. A., R. W. WILSON, "A measurement of excess antenna temperature at 4080 Mc/s", *Ap. J.*, **142**, 419.

PETTENGILL, G. H., R. B. DYCE, "Radar determination of the rotation of the planet Mercury", *Nature*, **206**, 1240.

SEEGER, C. L., G. WESTERHOUT, R. G. CONWAY, T. HOEKEMA, "A survey of the continuous radiation at a frequency of 400 Mc/s", *BAN*, **18**, 11.

TROITSKY, V. S., "Investigation of the surfaces of the Moon and planets by the thermal radiation", *Radio Science*, **69D**, 1585.

YAPLEE, V. S., S. H. KNOWLES, A. SHAPIRO, K. J. CRAIG, D. BROUWER, "The mean distance to the Moon as determined by radar", *Bull. Astron.*, **25**, 81.

WEINREB, S., *et al.*, "Observations of polarised OH emission", *Nature*, **208**, 440.

1966

ADGIE, R. L., H. GENT, "Positions of radio sources determined by the interferometer at the Royal Radar Establishment", *Nature*, **209**, 549.

BARRETT, A. H., A. E. E. ROGERS, "Observation of circularly polarised OH emission and narrow spectral features", *Nature*, **210**, 188.

DAVIES, R. D., G. DE JAGER, G. L. VERSCHUUR, "Detection of circular and linear polarisation in the OH emission features near W3 and W49", *Nature*, **209**, 974.

DENT, W. A., "Variation in the radio emission from the Seyfert galaxy NGC1275", *Ap. J.*, **144**, 843.

DENT, W. A., F. T. HADDOCK, "The extension of non-thermal radio source spectra to 8000 Mc/s", *Ap. J.*, **144,** 568.

HOWELL, T. F., J. R. SHAKESHAFT, "Measurement of the minimum cosmic background radiation at 20·7 cm wavelength", *Nature*, **210,** 1318.

KELLERMANN, K. I., "On the interpretation of radio source spectra and the evolution of radio galaxies and quasi-stellar sources", *Ap. J.*, **146,** 621.

LILLEY, A. E., P. PALMER, H. PENFIELD, B. ZUCKERMAN, "Radio astronomical detection of helium", *Nature*, **211,** 174.

LONGAIR, M. S., "On the interpretation of radio source counts", *MNRAS*, **133,** 421.

ROLL, P. G., and D. T. WILKINSON, "Cosmic background radiation at 3·2 cm—support for cosmic blackbody radiation", *Phys. Rev. Letters*, **16,** 405.

VÉRON, P., "Count of radio sources in the 3C revised catalogue", *Ann. Astrophys.*, **29,** 231.

WILLIAMS, P. J. S., "Magnetic field within some quasi-stellar radio sources", *Nature*, **210**, 285.

1967

ALLER, H. D., F. T. HADDOCK, "Time variations of the radio polarisation of quasi-stellar sources at 8000 Mc/s", *Ap. J.*, **147**, 833.

BARE, C., *et al.*, "Interferometer experiment with independent local oscillators", *Science*, **157,** 189.

BROTEN, N. W., *et al.*, "Observations of quasars using interferometer baselines up to 3074 km", *Nature*, **215,** 38.

COHEN, M. H., E. J. GUNDERMANN, "Angular diameter of 3C 279 from interplanetary scintillations", *Ap. J.*, **148,** L49.

DAVIES, R. D., *et al.*, "Measurements of OH emission sources with an interferometer of high resolution", *Nature*, **213,** 1109.

HEWISH, A., P. A. DENNISON, "Measurements of the solar wind and the small scale structure of the interplanetary medium", *J. Geophys. Res.*, **72,** 1977.

PALMER, H. P., H. GENT, *et al.*, "Radio diameter measurements with interferometer baselines of one and two million wavelengths", *Nature*, **213,** 789.

1968

GOLD, T., "Rotating neutron stars as the origin of the pulsating radio sources", *Nature*, **218,** 731.

HEWISH, A., S. J. BELL, J. D. PILKINGTON, P. F. SCOTT, R. A COLLINS, "Observation of a rapidly pulsating source", *Nature*, **217**, 709.

LARGE, M. I., A. E. VAUGHAN, B. Y. MILLS, "A pulsar supernova association", *Nature*, **220**, 340.

LYNE, A. G., AND F. G. SMITH, "Linear polarisation in pulsating radio sources", *Nature*, **218**, 124.

MORAN, J. M., *et al.*, "The structure of OH in W3", *Ap. J.*, **152,** 1197.

PAULINY-TOTH, I. I. K., K. I. KELLERMANN, "Repeated outbursts in the radio galaxy 3C 120", *Ap. J.*, **152,** L169.

PETTENGILL, G. H., T. W. THOMPSON, "A radar study of the lunar crater Tycho at 3·8 cm and 70 cm wavelength", *Icarus*, **8**, 457.

POOLEY, G. G., M. RYLE, "The extension of the number-flux density relation for radio sources to very small flux densities", *MNRAS*, **139**, 515.

STAELIN, D. H., E. C. REIFENSTEIN, "Pulsating radio sources near the Crab nebula", *Science*, **162**, 1481.

WILSON, W. J., A. H. BARRETT, "Discovery of hydroxyl radio emission from infrared stars", *Science*, **161**, 778.

1969

COCKE, W. J., M. J. DISNEY, D. J. TAYLOR, "Discovery of optical signals from the pulsar NPO532", *Nature*, **221**, 525.

COMELLA, J. H., H. D. CRAFT, R. V. E. LOVELACE, J. M. SUTTON, "Crab nebula pulsar NPO532", *Nature* **221**, 453.

KELLERMANN, K. I., I. I. K. PAULINY-TOTH, "The spectra of opaque radio sources", *Ap. J.*, **155**, L71.

MERKELIJN, J. K., "Accurate positions and optical identifications for 225 radio sources between declinations + 20° and − 33°", *Aust. J. Phys.*, **22**, 237.

MITTON, S., M. RYLE, "High resolution observations of Cygnus at 2·7 GHz and 5 GHz", *MNRAS*, **146**, 221.

OORT, J. H., "Infall of gas from intergalactic space", *Nature*, **224**, 1158.

POOLEY, G. G., "5C3: a radio continuum of M31 and its neighbourhood", *MNRAS*, **144**, 101.

1970

BOLTON, J. G., J. V. WALL, "Quasi-stellar objects in the Parkes 2700 MHz survey", *Aust. J. Phys.*, **23**, 789.

HEESCHEN, D. S., "Radio observations of E and SO galaxies", *Astron. J.*, **75**, 53.

KELLERMANN, K. I., *et al.*, "High resolution observations of compact radio sources at 13 cm", *Ap. J.*, **161**, 803.

WADE, C. M., "Precise positions of radio sources", *Ap. J.*, **162**, 381.

WILD, J. P., "Some investigations of the solar corona: the first two years of observations with Culgoora radioheliograph", *Proc. Astron. Soc. Aust.*, **1**, 365.

References to Source Surveys

Valuable guides to references of source surveys and lists of radio sources appear in the following books:

Appendix 3 in "*Radio Astronomy*" by J. D. Kraus (McGraw-Hill) 1966.
Appendix 4 in "*Radio Astrophysics*" by A. G. Pacholczyk (Freeman) 1970.

References to Radio Telescopes

Papers describing radio telescopes have appeared in a miscellany of journals. The following publications provide descriptions of radio telescopes and a guide to further references.

"*Radiotelescopes*" by W. N. Christiansen and J. A. Högbom (Cambridge) 1969.

Chapter 6 in "*Radio Astronomy*" by J. D. Kraus (McGraw-Hill) 1966.

Radio Astronomy Issue, *Proc. IRE* (*USA*), January 1958.

Radio Astronomy Issue, *Proc. IRE* (*Australia*), February 1963.

An account of the construction of the 250 ft radio telescope at Jodrell Bank is given in "*The Story of Jodrell Bank*" by A. C. B. Lovell (Oxford) 1968.

Books on Radio Astronomy

In addition to the many elementary and semipopular books on radio astronomy, the following are examples of notable books at a more advanced level.[1]

1954 *Meteor Astronomy*. A. C. B. Lovell (Oxford).

1955 *Radio Astronomy*. J. L. Pawsey and R. N. Bracewell (Oxford).

1957 *The Exploration of Space by Radio*. R. Hanbury Brown and A. C. B. Lovell (Chapman and Hall).

1957 *Radio Astronomy* (IAU Symposium No 4). Ed. H. C. van de Hulst (Cambridge).

1959 *Paris Symposium on Radio Astronomy* (IAU Symposium No 9). Ed. R. N. Bracewell (Stanford).

1960 *Cosmic Radio Waves*. I. S. Shklovsky (Transl. Russian edition 1956) (Harvard).

1961 *Meteor Science and Engineering*. D. W. R. McKinley (McGraw-Hill).

1963 *Radio Astronomy*. J. L. Steinberg and J. Lequeux (Transl. French edition 1960) McGraw-Hill.

1965 *Solar Radio Astronomy*, M. R. Kundu (Interscience).

1965 *Solar System Radio Astronomy*. Ed. J. Aarons (Plenum).

1966 *Radio Astronomy*. J. D. Kraus (McGraw-Hill).

1968 *Radar Astronomy*. Ed. J. V. Evans and T. Hagfors (McGraw-Hill).

1970 *Radio Emission of the Sun and Planets*. V. V. Zheleznyakov (Transl. Russian edition 1964) (Pergamon).

1970 *Radio Astrophysics*. A. G. Pacholczyk (Freeman).

[1] Many valuable review papers have also been published; a particularly useful series may be found in the Annual Reviews of Astronomy and Astrophysics (Annual Reviews Inc.).

Glossary

The following terms are explained as a guide to aid the general reader.

Frequency and Wavelength

Radio waves are electromagnetic waves of wavelength greater than 1 millimetre (mm). The frequency is the number of waves emitted per second. The name "Hertz" is now used for cycle per second, so that 1 Hz = 1 c/s. Frequencies in radio astronomy are usually stated in MHz (= megahertz = million cycles per second).

A simple formula for converting wavelength (λ) in metres to frequency (f) in MHz is $f = 300/\lambda$.

Power Flux Density

Received power is measured in terms of the power flowing across 1 sq. metre (m^2) per cycle/sec (Hz) of received bandwidth. The units of power flux density are therefore watts per sq. metre per Hz, usually written $W\ m^{-2}\ Hz^{-1}$.

As radio astronomy power flux densities are extremely small, a convenient unit is $10^{-26}\ W\ m^{-2}\ Hz^{-1}$ and is commonly referred to as the flux unit (f.u.).

Radio Noise

Rapid random variations of amplitude are a normal characteristic of naturally produced radio emission from astronomical sources or of the radio power generated within the receiver by the irregular motion of electrons in the radio components. The radiation is described as radio noise, and the receiver is designed to measure the mean power.

Radio Brightness Temperature

Thermal emission extends to radio wavelengths and provides a comparatively weak radiation according to Planck's radiation law, which at radio wavelengths can be expressed in a simpler form known as the Rayleigh-Jeans law. The emission is proportional to the temperature, and for a dense source the power falls off inversely as λ^2, where λ is the wavelength. In contrast, non-thermal processes can produce very strong radiation. The intensity of non-thermal radio emission can also be expressed in terms of an effective temperature, which is the temperature that would be required in order to produce the same radio intensity by thermal emission. This effective (or equivalent) temperature, is known as the "radio brightness temperature".

Radio Telescope Beamwidth and Resolution

A radio telescope is an aerial (or antenna) used for radio astronomy.

The beamwidth is the angle over which radio waves are received. The beamwidth is normally measured between the directions of half the maximum power sensitivity. The larger the radio telescope and the shorter the wavelength the narrower is the beamwidth. For a radio telescope of aperture diameter D, the beamwidth is approximately λ/D radians, or about 60 λ/D degrees. Two sources cannot be resolved (observed separately) if they are closer together than the beamwidth.

The term 'radio telescope' may be assumed to include the radio receiving equipment. In a similar way, the name 'radiometer' which is often applied to the receiver and recorder only should more properly be used to describe a complete aerial and receiving system.

Polar Diagram and Sidelobes

A certain amount of radio power can be received in directions outside the main beam and these are known as sidelobes. The sensitivity of a radio telescope in different directions including the main beam is often shown graphically by the 'aerial polar diagram'.

Interferometer Pattern

The polar diagram of two separated aerials joined together is a pattern of maxima and minima known as interferometer lobes. The maxima correspond to directions such that the waves at the two aerials combine together in the same phase (in step); the minima occur when the waves are out-of-step.

An interferometer of spacing L between the aerial produces lobes of angular separation λ/L radians (about 60 λ/L degrees).

A simple interferometer records the background radiation as well as discrete sources. The phase-switching interferometer is a modification to eliminate the background. If the signal from one aerial is reversed in phase, the maxima and minima of the interferometer lobes are interchanged. If now the difference is measured, the background output is eliminated and only discrete sources record a response. (*See* Figure 4.1.)

Spatial Frequency and Fourier Analysis

The number of interferometer lobes per unit angle (radian) is called the spatial frequency. Observing a complex source in the sky with a given interferometer spacing provides the response at one spatial frequency. Any source distribution can be regarded as a combination of many different spatial frequencies, and so can be derived from interferometer observations at many different spacings. This is the principle of Fourier synthesis.

Linear Polarisation and Faraday Rotation

If the electric field in a radio wave is confined to one direction the wave is described as linearly polarised. If the wave passes through an ionised gas in the presence of a magnetic field, the direction of polarisation alters. This is known as Faraday rotation of the polarisation because the effect, in principle, was first discovered in light waves by Faraday.

Circular and Random Polarisation

If a radio wave is emitted with a continually-rotating direction of polarisation it is said to be circularly polarised. If a source emits radio waves with polarisation of random orientation it is said to be randomly polarised.

Plasma Frequency

In an ionised gas, the electrons have a natural resonant frequency of oscillation called the plasma frequency. The more electrons, the higher the plasma frequency. A radio wave is totally reflected where the electron density is so great that the plasma frequency equals or exceeds the radio frequency. Consequently, lower frequencies (or longer wavelengths) are more easily reflected or refracted by ionised atmospheres. (The precise relations depend not only on the electron density, but also on the magnetic field, if present, and angle of incidence).

Absorption

Absorption of a radio wave occurs when the vibrating electrons collide with positive ions, and consequently increases with density. Absorption can be particularly severe when the electron density is such that the resonant plasma frequency is close to the radio frequency.

Gyro Frequency

When a magnetic field is present, electrons spiral around the magnetic field with a frequency of gyration equal to $2 \cdot 8\ B$ in MHz, where B is the magnetic field in gauss.

Right Ascension and Declination

In describing the directions of astronomical objects it is useful to imagine the sky as a celestial sphere spinning about the N–S axis of the Earth.

The Declination (Dec.) of a source is reckoned in degrees N (+) or S (−) from the celestial equator.

The Right Ascension (R.A.) of a source is measured by the difference in time (in hours and minutes) when the source crosses the meridian compared with that of a reference direction known as the first point of Aries.

Galactic Coordinates

Positions in the Galaxy are often stated in galactic latitude and longitude with respect to the galactic plane and centre. The system was revised in 1960, and the centre or 0° on the new system corresponds to longitude 327°69, latitude − 1°40 in the old coordinates.

Distances

Radio waves in free space travel with the velocity of light, approximately 300 000 kilometres per sec (3×10^5 km/sec).

Large distances are measured in light-years, the distance travelled by light in one year.

1 light-year = $9 \cdot 45 \times 10^{12}$ km

Other units used in astronomy are:

(a) The astronomical unit (AU) is the mean radius of the Earth's orbit round the Sun.

1 AU = 149 600 000 km = $1 \cdot 5 \times 10^{8}$ km approx.

(b) The parsec (pc) is the distance at which the mean radius of the Earth's orbit subtends 1 second of arc.

1 parsec = $3 \cdot 26$ light years

$\doteq 3 \times 10^{13}$km = 2×10^{5} AU approx.

Redshift

According to Hubble's law, the light from very distant objects is shifted towards longer wavelengths by an amount proportional to the distance. The fractional increase in wavelength ($\delta\lambda/\lambda$) is called the redshift Z.

The factor by which the distance is calculated from the redshift tends to be subject to revision as new astronomical data becomes available.

Hydrogen Line Profile

The spectral line at a wavelength of about 21 cm is caused by a hyperfine transition in neutral atomic hydrogen. The frequency of the line measured in the laboratory has been found to be 1420·405 MHz.

In radio astronomical observations, the relative motions of hydrogen in interstellar space produce shifts in the observed frequency owing to the Doppler effect. Consequently the observed recording may cover a considerable bandwidth, and the intensity plotted against frequency is described as the line profile.

The astronomical importance of the 21 cm hydrogen line has been recognized by international agreement, and to ensure maximum protection from interference the band from 1400 to 1427 MHz has, since 1959, been exclusively allocated to radio astronomy.

Subject Index

Name Index